Mathematics for Technicians

Scotec Edition

Level 2 Electrical Engineering Mathematics

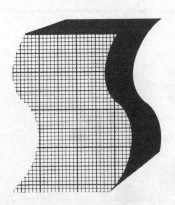

Mathematics for Technicians

Second Edition

Level 2
Electrical Engineering
Mathematics

Mathematics for Technicians

Scotec Edition

Level 2 Electrical Engineering Mathematics

R. B. Buchan

B.Sc., F.S.S., A.F.I.M.A.,
Head of Mathematics
Esk Valley College

A. Greer

C.Eng., M.R.Ae.S.
Senior Lecturer
City of Gloucester College of Technology

G. W. Taylor

B.Sc.(Eng.), C.Eng., M.I.Mech.E.
Principal Lecturer
City of Gloucester College of Technology

Stanley Thornes (Publishers) Ltd.

First published in 1980 by:
Stanley Thornes (Publishers) Ltd
EDUCA House
Liddington Estate
Leckhampton Road
CHELTENHAM GL53 0DN
England

British Library Cataloguing in Publication Data
Buchan, R B
    Electrical engineering mathematics for SCOTEC
    Level II. – (Mathematics for technicians series).
    1. Electric engineering – Mathematics
    I. Title   II. Greer, Alec   III. Taylor,
    Graham William   IV. Series
    510′.2′46213   TK153

    ISBN 0–85950–477–8

Typeset in Monotype 10/12 Times New Roman by
Gloucester Typesetting Co Ltd
Printed and bound in Great Britain by
The Pitman Press, Bath

CONTENTS

AUTHORS' NOTE ON THE SERIES

Arising from the recommendations in the Hudson Report, with particular reference to the Haycocks Report, the Scottish Technical Education Council was set up in 1973. The Council have established six Sector Committees each responsible for a group of courses.

A major change of emphasis in the educational approach adopted in each SCOTEC course has been introduced by the use of 'objectives' in the syllabus which allows student and lecturer to achieve planned progress through the course on a step-by-step basis.

This set of books started with the Introductory Course which provides the mathematics required by Level I Technicians and continues at Level II with Engineering Mathematics (this volume), together with the sister book 'Elements of Engineering Mechanics'. Each book follows a standard pattern, and each chapter starts with the words "After reading the end of this chapter you should be able to:" and this statement is followed by the objectives for that particular topic as laid down in the syllabus. Thereafter each chapter contains explanatory text, worked examples and copious supplies of further exercises. As planned at present the series comprises:-

AN INTRODUCTORY COURSE   Level I
MECHANICAL ENGINEERING
MATHEMATICS   Level II
ELECTRICAL ENGINEERING
MATHEMATICS   Level II
MATHEMATICS FOR
ENGINEERING TECHNICIANS   Level III

R. E. Burden
A. Cross
G. W. Taylor

1980

SIMULTANEOUS LINEAR EQUATIONS

1.□

After reaching the end of this chapter you should be able to:
Solve a pair of simultaneous equations by
substitution and by elimination.

SOLUTION OF SIMULTANEOUS LINEAR EQUATIONS

Consider the equations:

$$3x+2y = 7$$

$$4x+y = 6$$

The unknown quantities x and y appear in both equations. To solve the equations we have to find values of x and y so that *both* equations are satisfied. Such equations are called *simultaneous equations*.

Three methods are available for solving simultaneous equations.

1. Substitution Method

EXAMPLE 1

Solve the equations:

$$2x+y = 10 \qquad (1)$$

$$3x+2y = 17 \qquad (2)$$

We can write equation (1) above as:

$$y = 10-2x \qquad (3)$$

and, substituting this value of y into equation (2), we have:

$$3x+2(10-2x) = 17$$

and we now have an equation with x the only unknown.

$$\therefore \qquad 3x+20-4x = 17$$

$$\therefore \qquad x = 3$$

1

Substituting this value for x in equation (3),

$$y = 10 - 2(3)$$

$$\therefore \qquad y = 4$$

The solutions are therefore,

$$x = 3 \quad \text{and} \quad y = 4$$

The solutions should always be checked by substituting the values found into each of the original equations:

Equation (1) has:

$$\text{L.H.S.} = 2(3)+4 = 10 = \text{R.H.S.}$$

and equation (2) has:

$$\text{L.H.S.} = 3(3)+2(4) = 17 = \text{R.H.S.}$$

EXAMPLE 2

Solve the equations:

$$2x+3y = 16 \tag{1}$$

$$3x+2y = 14 \tag{2}$$

From equation (1):

$$3y = 16 - 2x$$

$$\therefore \qquad y = \frac{16-2x}{3} \tag{3}$$

Substituting this value in equation (2),

$$3x + \frac{2(16-2x)}{3} = 14$$

and, multiplying through by 3, we get:

$$9x + 2(16-2x) = 42$$

from which $\qquad x = 2$

Substituting this value for x in equation (3), we have:

$$y = \frac{16-2(2)}{3} = \frac{16-4}{3}$$

$$\therefore \qquad y = 4$$

The solutions are therefore:

$$x = 2 \quad \text{and} \quad y = 4$$

Checking these values by substituting into the original equations we have:

Equation (1) has:

$$\text{L.H.S.} = 2(2)+3(4) = 16 = \text{R.H.S.}$$

Equation (2) has:

$$\text{L.H.S.} = 3(2)+2(4) = 14 = \text{R.H.S.}$$

2. Elimination Method

This method is most generally used in solving equations which contain the first power only of the unknown quantities.

EXAMPLE 3

Solve the equations:

$$3x+4y = 11 \tag{1}$$

$$x+7y = 15 \tag{2}$$

If we multiply equation (2) by 3 we shall have the same coefficient of x in each of the equations:

$$3x+21y = 45 \tag{3}$$

We can now eliminate x by subtracting equation (1) from equation (3).

$$(3x+21y)-(3x+4y) = 45-11$$

$$\therefore \qquad 17y = 34$$

$$\therefore \qquad y = 2$$

To find x we may substitute in either of original equations.

Substituting in equation (1):

$$3x+4(2) = 11$$

$$\therefore \qquad x = 1$$

Therefore the solutions are:

$$x = 1 \quad \text{and} \quad y = 2$$

To check these values substitute them in equation (2). (There would be no point in substituting them in equation (1) for this was used in finding x from the y value.) Substituting in equation (2),

$$\text{L.H.S.} = 1+7(2) = 15 = \text{R.H.S.}$$

EXAMPLE 4

Solve the equations:

$$5x+3y = 29 \tag{1}$$

$$4x+7y = 37 \tag{2}$$

The same coefficient of x can be obtained if equation (1) is multiplied by 4 and equation (2) by 5. As before, we may then subtract and x will disappear.

Multiplying equation (1) by 4,

$$20x+12y = 116 \tag{3}$$

Multiplying equation (2) by 5,

$$20x+35y = 185 \tag{4}$$

Subtracting equation (3) from equation (4),

$$(35-12)y = 185-116$$

$$\therefore \qquad y = 3$$

Substituting in equation (1),

$$5x+3(3) = 29$$

$$\therefore \qquad x = 4$$

Therefore the solutions are:

$$x = 4 \quad \text{and} \quad y = 3$$

A check on these values is made by substituting them into equation (2):

$$\text{L.H.S.} = 4(4)+7(3) = 37 = \text{R.H.S.}$$

Frequently, in practice, the coefficients of the unknowns are not whole numbers. The same methods apply but care must be taken with the arithmetic.

EXAMPLE 5

Solve the equations:

$$3.175x+0.238y = 6.966 \tag{1}$$

$$2.873x+4.192y = 11.804 \tag{2}$$

To eliminate, say, x we must arrange for x to have the same coefficient in both equations. To achieve this we multiply equation (1) by the coefficient of x in equation (2) and then equation (2) by the coefficient of x in equation (1).

Multiplying equation (1) by 2.873,

$$9.122x + 0.683\,8y = 20.02 \tag{3}$$

Multiplying equation (2) by 3.175,

$$9.122x + 13.31y = 37.48 \tag{4}$$

Subtracting equation (3) from equation (4),

$$12.63y = 17.46$$

$$\therefore \qquad y = 1.383$$

Substituting this value in equation (1),

$$3.175x + 0.238(1.383) = 6.966$$

$$\therefore \qquad x = \frac{6.966 - 0.329\,7}{3.175}$$

$$\therefore \qquad x = 2.089$$

Therefore the solutions are:

$$x = 2.089 \quad \text{and} \quad y = 1.383$$

A check on these values may be made by substituting them into equation (2):

L.H.S. $= 2.873(2.089) + 4.192(1.383) = 11.804 =$ R.H.S.

EXAMPLE 6

Solve the equations:

$$\frac{2x}{3} - \frac{y}{4} = \frac{7}{12} \tag{1}$$

$$\frac{3x}{4} - \frac{2y}{5} = \frac{3}{10} \tag{2}$$

In this example it is best to clear each equation of fractions before attempting to solve simultaneously.

Multiplying equation (1) by 12,

$$8x - 3y = 7 \tag{3}$$

Multiplying equation (2) by 20,

$$15x - 8y = 6 \tag{4}$$

We can now proceed in the usual way.

Multiplying equation (4) by 8,

$$120x - 64y = 48 \tag{5}$$

Multiplying equation (3) by 15,

$$120x - 45y = 105 \qquad (6)$$

Subtracting equation (5) from equation (6),

$$-45y - (-64)y = 105 - 48$$

$$\therefore \qquad y = 3$$

Substituting in equation (3),

$$8x - 3(3) = 7$$

$$\therefore \qquad x = 2$$

Hence solution is:

$$x = 2 \quad \text{and} \quad y = 3$$

A check on these values will necessitate substitution into both equations (1) and (2) since both were modified before any elimination took place:

Equation (1) has:

$$\text{L.H.S.} = \frac{2(2)}{3} - \frac{3}{4} = \frac{7}{12} = \text{R.H.S.}$$

Equation (2) has:

$$\text{L.H.S.} = \frac{3(2)}{4} - \frac{2(3)}{5} = \frac{3}{10} = \text{R.H.S.}$$

3. Graphical Method

Simultaneous linear equations may be solved by plotting the graphs of the two equations and finding where they intersect. This method is explained fully in Chapter 7.

PROBLEMS INVOLVING SIMULTANEOUS EQUATIONS

In problems which involve two unknowns it is necessary to form two separate equations from the given data and then to solve these as shown above.

EXAMPLE 7

In a certain lifting machine it is found that the effort (E) and the load (W) which is being raised are connected by the equation $E = aW + b$. An effort of 3.7 units raises a load of 10 units whilst an effort of 7.2 units raises a load of 20 units. Find the values of the constants a and b and hence find the effort needed to lift a load of 12 units.

Substituting $E = 3.7$ and $W = 10$ into the given equation we have,

$$3.7 = 10a + b \tag{1}$$

Substituting $E = 7.2$ and $W = 20$ into the given equation we have,

$$7.2 = 20a + b \tag{2}$$

Subtracting equation (1) from equation (2),

$$3.5 = 10a$$

$$a = 0.35$$

Substituting for a in equation (1),

$$3.7 = 10 \times 0.35 + b$$

$\therefore$
$$3.7 = 3.5 + b$$

$\therefore$
$$3.7 - 3.5 = b$$

$\therefore$
$$b = 0.2$$

The given equation therefore becomes:

$$E = 0.35W + 0.2$$

When
$$W = 12,$$

then
$$E = 0.35 \times 12 + 0.2 = 4.2 + 0.2 = 4.4 \text{ units}$$

Hence an effort of 4.4 units is needed to raise a load of 12 units.

EXAMPLE 8

The currents I_1 and I_2 in a certain circuit are connected by the following equations:

$$0.4I_1 - 0.3I_2 = 3 \tag{1}$$

$$1.1I_1 - 0.2I_2 = 5 \tag{2}$$

Find I_1 and I_2.

Multiplying equation (1) by 1.1 we get,

$$0.44I_1 - 0.33I_2 = 3.3 \tag{3}$$

Multiplying equation (2) by 0.4 we get,

$$0.44I_1 - 0.08I_2 = 2.0 \tag{4}$$

Subtracting equation (4) from equation (3),

$$-0.25I_2 = 1.3$$

$$\therefore \qquad I_2 = \frac{1.3}{-0.25}$$

$$\therefore \qquad I_2 = -5.2$$

Substituting for I_2 in equation (1) we get,

$$0.4I_1 - 0.3 \times (-5.2) = 3$$

$$\therefore \qquad 0.4I_1 + 1.56 = 3$$

$$0.4I_1 = 3 - 1.56$$

$$\therefore \qquad 0.4I_1 = 1.44$$

$$\therefore \qquad I_1 = \frac{1.44}{0.4}$$

$$\therefore \qquad I_1 = 3.6$$

EXAMPLE 9

Two equations connecting resistances R_1 and R_2 in an electric circuit are:

$$\frac{3}{R_1} + \frac{4}{R_2} = 1.6$$

$$\frac{5}{R_1} + \frac{8}{R_2} = 3.0$$

Find the values of R_1 and R_2.

Let $x = \dfrac{1}{R_1}$ and $y = \dfrac{1}{R_2}$, then,

$$3x + 4y = 1.6 \qquad (1)$$

$$5x + 8y = 3.0 \qquad (2)$$

Multiplying (1) by 2 $\qquad 6x + 8y = 3.2 \qquad (3)$

Subtracting (2) from (3), $\qquad x = 0.2$

Substituting for x in (2), $\qquad 5 \times 0.2 + 8y = 3.0$

$$y = 0.25$$

$$\therefore \qquad R_1 = \frac{1}{0.2} = 5$$

and, $\qquad R_2 = \dfrac{1}{0.25} = 4$

EXAMPLE 10

A heating installation for one house consists of 5 radiators and 4 convector heaters and the cost of the installation is £270. In a second house 6 radiators and 7 convector heaters are used and the cost of this installation is £402. In each house the installation costs are £50. Find the cost of a radiator and the cost of a convector heater.

For the first house the cost of the hardware is:

$$£270 - £50 = £220$$

For the second house the cost of the hardware is:

$$£402 - £50 = £352$$

Let $£x$ be the cost of a radiator and $£y$ the cost of a convector heater.

For the first house, $5x+4y = 220$ (1)

For the second house, $6x+7y = 352$ (2)

Multiplying (1) by 6: $30x+24y = 1320$, and   (3)

Multiplying (2) by 5: $30x+35y = 1760$ (4)

Subtracting equation (3) from equation (4) then,

$$11y = 440$$

$\therefore$ $y = 40$

and substituting for $y = 40$ in equation (1) then,

$$5x+4(40) = 220$$

$\therefore$ $5x = 60$

$\therefore$ $x = 12$

Therefore the cost of a radiator is £12 and the cost of a convector heater is £40.

Exercise 1

Solve the following simultaneous equations:

1) $3x+2y = 14$
   $2x+5y = 24$

2) $7x-3y = -2$
   $8x-2y = 2$

3) $\dfrac{x}{4}+\dfrac{y}{5} = \dfrac{3}{2}$

   $2x+3y = 19$

4) $\dfrac{x}{2}+\dfrac{y}{3} = \dfrac{13}{6}$

   $\dfrac{2x}{7}-\dfrac{y}{4} = \dfrac{5}{14}$

2. QUADRATIC EQUATIONS

After reaching the end of this chapter you should be able to:-

1. Define the roots of an equation.
2. Determine the equation which is satisfied by a given pair of roots.
3. Define a quadratic expression and a quadratic equation.
4. Add a constant term to an expression such as ax^2+bx to make it a perfect square.
5. Develop from (4) the formula for solving a quadratic equation.
6. Solve simple quadratic equations, which provide real roots, by the use of the formula.
7. Form and solve quadratic equations which are mathematical models of phsyical problems.

ROOTS OF AN EQUATION

If either of two factors has zero value, then their product is zero. Thus if either $M = 0$ or $N = 0$ then $M \times N = 0$.

Now suppose that either $\qquad x = 1$ or $\qquad x = 2$

$\therefore$ rearranging either $\qquad x - 1 = 0$ or $x - 2 = 0$

Hence: $\qquad (x-1)(x-2) = 0$

since either of the factors has zero value.

If we now multiply out the brackets of this equation we have:

$$x^2 - 3x + 2 = 0$$

and we know that $x = 1$ and $x = 2$ are values of x which satisfy this equation. The values 1 and 2 are called the solutions or *roots* of the equation $x^2 - 3x + 2 = 0$.

EXAMPLE 1

Find the equation whose roots are -2 and 4.

From the values given either $\qquad x = -2 \qquad$ or $\qquad x = 4$

$\therefore \qquad$ either $\quad x + 2 = 0 \qquad$ or $\quad x - 4 = 0$

Hence: $\qquad (x+2)(x-4) = 0$

since either of the factors has zero value.

$\therefore$ Multiplying out: $\quad x^2 - 2x - 8 = 0$

12

EXAMPLE 2

Find the equation whose roots are 3 and −3.

From the values given either $\qquad x = 3$ or $\qquad x = -3$

∴ $\qquad\qquad\qquad$ either $x - 3 = 0$ or $x + 3 = 0$

Hence: $\qquad\qquad\qquad (x-3)(x+3) = 0$

since either of the factors has zero value.

Multiplying out we have:

$$x^2 - 9 = 0$$

EXAMPLE 3

Find the equation whose roots are 5 and 0.

From the values given either $\qquad x = 5$ or $x = 0$

∴ $\qquad\qquad\qquad$ either $x - 5 = 0$ or $x = 0$

Hence: $\qquad\qquad\qquad x(x-5) = 0$

since either of the factors has zero value.

and multiplying out we have:

$$x^2 - 5x = 0$$

Exercise 2

Find the equations whose roots are:

1) 3, 1 $\qquad$ **5)** 2.73, −1.66 $\qquad$ **9)** −3.5, +3.5

2) 2, −4 $\qquad$ **6)** −4.76, −2.56 $\qquad$ **10)** repeated, each = 4

3) −1, −2 $\qquad$ **7)** 0, 1.4

4) 1.6, 0.7 $\qquad$ **8)** −4.36, 0

QUADRATIC EQUATIONS

An equation of the type $ax^2 + bx + c = 0$, involving x in the second degree and containing no higher power of x, is called a *quadratic equation*. The constants a, b and c can have any numerical values. Thus,

$$x^2 - 9 = 0 \quad \text{where } a = 1, b = 0 \text{ and } c = -9,$$

$$x^2 - 2x - 8 = 0 \quad \text{where } a = 1, b = -2 \text{ and } c = -8,$$

$$2.5x^2 - 3.1x - 2 = 0 \quad \text{where } a = 2.5, b = -3.1 \text{ and } c = -2,$$

are all examples of quadratic equations. A quadratic equation may contain only the square of the unknown quantity, as in the first of the above equations, or it may contain both the square and the first power as in the other two.

Three methods are available for solving quadratic equations.

1. Solution by Factors

This method is the reverse of the procedure used to find an equation when given the roots. We shall now start with the equation and proceed to solve the equation and find the roots.

We shall again use the fact that if the product of two factors is zero then one factor or the other must be zero. Thus if $M \times N = 0$ then either $M = 0$ or $N = 0$.

When the factors are easy to find the factor method is very quick and simple. However do not spend too long trying to find factors: if they are not easily found use the formula given in the next method (page 16) to solve the equation.

EXAMPLE 4

Solve the equation $(2x+3)(x-5) = 0$.

Since the product of the two factors $2x+3$ and $x-5$ is zero then either

$$2x+3 = 0 \quad \text{or} \quad x-5 = 0$$

Hence:
$$x = -\frac{3}{2} \quad \text{or} \quad x = 5$$

EXAMPLE 5

Solve the equation $6x^2+x-15 = 0$.

Factorising, $(2x-3)(3x+5) = 0$

$\therefore$ either $2x-3 = 0 \quad \text{or} \quad 3x+5 = 0$

Hence:
$$x = \frac{3}{2} \quad \text{or} \quad x = -\frac{5}{3}$$

EXAMPLE 6

Solve the equation $14x^2 = 29x - 12$.

Bring all the terms to the left-hand side:

$$14x^2 - 29x + 12 = 0$$

$$\therefore \qquad (7x-4)(2x-3) = 0$$

$$\therefore \qquad \text{either} \quad 7x-4 = 0 \quad \text{or} \quad 2x-3 = 0$$

Hence: $\qquad\qquad\qquad x = \dfrac{4}{7} \quad \text{or} \quad x = \dfrac{3}{2}$

EXAMPLE 7

Find the roots of the equation $x^2 - 16 = 0$.

Factorising, $\qquad\qquad\qquad (x-4)(x+4) = 0$

$$\therefore \qquad\qquad \text{either} \quad x-4 = 0 \quad \text{or} \quad x+4 = 0$$

Hence: $\qquad\qquad\qquad\qquad x = 4 \quad \text{or} \quad x = -4$

In this case an alternative method may be used:

Rearranging the given equation: $x^2 = 16$

and taking the square root of both sides: $x = \sqrt{16} = \pm 4$

Remember that when we take a square root we must insert the $\pm$ sign, because $(+4)^2 = 16$ and $(-4)^2 = 16$.

EXAMPLE 8

Solve the equation $x^2 - 2x = 0$.

Factorising, $\qquad\qquad\qquad x(x-2) = 0$

$$\therefore \qquad\qquad \text{either} \quad x = 0 \quad \text{or} \quad x-2 = 0$$

Hence: $\qquad\qquad\qquad\qquad x = 0 \quad \text{or} \quad x = 2$

Note: the solution $x = 0$ must not be omitted as it is a solution in the same way as $x = 2$ is a solution. Equations should not be divided through by variables, such as x, since this removes a root of the equation.

EXAMPLE 9

Solve the equation $x^2 - 6x + 9 = 0$.

Factorising, $\qquad\qquad\qquad (x-3)(x-3) = 0$

$$\therefore \qquad\qquad \text{either} \quad x-3 = 0 \quad \text{or} \quad x-3 = 0$$

Hence: $\qquad\qquad\qquad\qquad x = 3 \quad \text{or} \quad x = 3$

In this case there is only one arithmetical value for the solution. Technically, however, there are two roots and when they have the same numerical value they are said to be repeated roots.

2. Solution by Formula

In general quadratic expressions do not factorise and therefore some other method of solving quadratic equations must be used.

Consider the expression $ax^2+bx = a\left(x^2+\dfrac{b}{a}x.\right)$

If we add (half the coefficient of x^2) to the terms inside the bracket we get

$$ax^2+bx = a\left[x^2+\frac{b}{a}x+\left(\frac{b}{2a}\right)^2\right]-a.\left(\frac{b}{2a}\right)^2$$

$$= a\left(x+\frac{b}{2a}\right)^2-\frac{b^2}{4a}$$

We are said to have completed the square of ax^2+bx,

We shall now establish a formula which may be used to solve any quadratic equation. If:

$$ax^2+bx+c = 0$$

$$ax^2+bx = -c$$

Completing the square of the L.H.S.,

$$a\left(x+\frac{b}{2a}\right)^2-\frac{b^2}{4a} = -c$$

$$4a^2\left(x+\frac{b}{2a}\right)^2-b^2 = -4ac$$

$$4a^2\left(x+\frac{b}{2a}\right)^2 = b^2-4ac$$

Taking the square root of both sides:

$$2a\left(x+\frac{b}{2a}\right)= \pm\sqrt{b^2-4ac}$$

$$x = \frac{-b\pm\sqrt{b^2-4ac}}{2a}$$

The *standard form* of the *quadratic equation* is:

$$\boxed{ax^2+bx+c = 0}$$

As shown above the *solution* of this equation is:

$$\boxed{x = \frac{-b\pm\sqrt{b^2-4ac}}{2a}}$$

EXAMPLE 10

Solve the equation $3x^2 - 8x + 2 = 0$.

Comparing with $ax^2 + bx + c = 0$, we have $a = 3$, $b = -8$ and $c = 2$.

Substituting these values in the formula,

$$x = \frac{-(-8) \pm \sqrt{(-8)^2 - 4 \times 3 \times 2}}{2 \times 3}$$

$$= \frac{8 \pm \sqrt{64 - 24}}{6}$$

$$= \frac{8 \pm \sqrt{40}}{6}$$

$$= \frac{8 \pm 6.325}{6}$$

$\therefore$ either $x = \dfrac{8 + 6.325}{6}$ or $x = \dfrac{8 - 6.325}{6}$

Hence:

$$x = 2.39 \quad \text{or} \quad x = 0.28$$

EXAMPLE 11

Solve the equation $2.13x^2 + 0.75 - 6.89 = 0$.

Here $a = 2.13$, $b = 0.75$, $c = -6.89$.

$$x = \frac{-0.75 \pm \sqrt{(0.75)^2 - 4(2.13)(-6.89)}}{2 \times 2.13}$$

$$= \frac{-0.75 \pm \sqrt{0.562\,5 + 58.70}}{4.26}$$

$$= \frac{-0.75 \pm \sqrt{59.26}}{4.26}$$

$$= \frac{-0.75 \pm 7.698}{4.26}$$

$\therefore$ either $x = \dfrac{-0.75 + 7.698}{4.26}$ or $x = \dfrac{-0.75 - 7.698}{4.26}$

Hence

$$x = 1.631 \quad \text{or} \quad x = -1.983$$

EXAMPLE 12

Solve the equation $-2x^2+3x+7 = 0$.

Where the coefficient of x^2 is negative it is best to make it positive by multiplying both sides of the equation by (-1). This is equivalent to changing the sign of each of the terms. Thus:

$$2x^2-3x-7 = 0$$

This gives $a = 2$, $b = -3$ and $c = -7$.

$$\therefore \qquad x = \frac{-(-3)\pm\sqrt{(-3)^2-4\times2\times(-7)}}{2\times2}$$

$$= \frac{3\pm\sqrt{9+56}}{4} = \frac{3\pm\sqrt{65}}{4}$$

$$= \frac{3\pm8.063}{4}$$

$\therefore \qquad$ either $\quad x = \dfrac{3+8.063}{4} \quad$ or $\quad x = \dfrac{3-8.063}{4}$

Hence

$$x = 2.766 \qquad \text{or} \quad x = -1.266$$

EXAMPLE 13

Solve the equation $\dfrac{3}{2x-3}-\dfrac{2}{x+1} = 5$.

The L.C.M. of the denominators is $(2x-3)(x+1)$. Multiplying both sides of the equation by this gives:

$$3(x+1)-2(2x-3) = 5(2x-3)(x+1)$$

$$\therefore \qquad 3x+3-4x+6 = 5(2x^2-x-3)$$

$$\therefore \qquad -x+9 = 10x^2-5x-15$$

$$\therefore \qquad 10x^2-4x-24 = 0$$

Here $a = 10$, $b = -4$ and $c = -24$.

$$\therefore \qquad x = \frac{-(-4)\pm\sqrt{-(4)^2-4(10)(-24)}}{2\times10}$$

$$= \frac{4\pm\sqrt{16+960}}{20}$$

$$= \frac{4\pm\sqrt{976}}{20} = \frac{4\pm31.24}{20}$$

$\therefore$ either $x = \dfrac{4+31.24}{20}$ or $x = \dfrac{4-31.24}{20}$

Hence:

$$x = 1.762 \qquad \text{or} \quad x = -1.362$$

EXAMPLE 14

Solve the equation $x^2+4x+5 = 0$.

Here $a = 1$, $b = 4$ and $c = 5$.

$\therefore$
$$x = \dfrac{-4\pm\sqrt{4^2-4(1)(5)}}{2(1)}$$

$$= \dfrac{-4\pm\sqrt{16-20}}{2}$$

$$= \dfrac{-4\pm\sqrt{-4}}{2}$$

Now when a number is squared the answer must be a positive quantity because two quantities having the same sign are being multiplied together. Therefore the square root of a negative quantity, as $\sqrt{-4}$ in the above equation, has no arithmetical meaning and is called an imaginary quantity. The equation $x^2+4x+5 = 0$ is said to have imaginary or complex roots. Equations which have complex roots are beyond the scope of this book and are dealt with in more advanced mathematics.

3. Graphical Method

Quadratic equations may be solved by plotting graphs. This method is explained fully in Chapter 7.

Exercise 3

Solve the following equations by the factor method:

1) $x^2-36 = 0$

2) $4x^2-6.25 = 0$

3) $9x^2-16 = 0$

4) $x^2+9x+20 = 0$

5) $x^2+x-72 = 0$

6) $3x^2-7x+2 = 0$

7) $m^2 = 6m-9$

8) $m^2+4m+4 = 36$

9) $14q^2 = 29q-12$

10) $9x+28 = 9x^2$

Solve the following equations by using the quadratic formula:

11) $4x^2-3x-2 = 0$

12) $x^2-x+\frac{1}{4} = \frac{1}{9}$

13) $3x^2+7x-5 = 0$ **16)** $2x^2-7x = 3$

14) $7x^2+8x-2 = 0$ **17)** $x^2+0.3x-1.2 = 0$

15) $5x^2-4x-1 = 0$ **18)** $2x^2-5.3x+1.25 = 0$

Solve the following equations:

19) $x(x+4)+2x(x+3) = 5$ **23)** $\dfrac{6}{x}-2x = 2$

20) $x^2-2x(x-3) = -20$ **24)** $40 = \dfrac{x^2}{80}+4$

21) $\dfrac{2}{x+2}+\dfrac{3}{x+1} = 5$ **25)** $\dfrac{x+2}{x-2} = x-3$

22) $\dfrac{x+2}{3}-\dfrac{5}{x+2} = 4$ **26)** $\dfrac{1}{x+1}-\dfrac{1}{x+3} = 15$

PROBLEMS INVOLVING QUADRATIC EQUATIONS

EXAMPLE 15

The diagonal of a rectangle is 15 cm long and one side is 2 cm longer than the other. Find the dimensions of the rectangle.

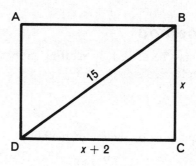

Fig. 2.1

In Fig. 2.1, let the length of BC be x cm. The length of CD is then $(x+2)$ cm. $\triangle$ BCD is right-angled and so by Pythagoras,

$$x^2+(x+2)^2 = 15^2$$

$$x^2+x^2+4x+4 = 225$$

$$2x^2+4x-221 = 0$$

Here $a = 2$, $b = 4$ and $c = -221$

$$\therefore \qquad x = \frac{-4 \pm \sqrt{4^2 - 4 \times 2 \times (-221)}}{2 \times 2}$$

$$\therefore \qquad x = 9.56 \quad \text{or} \quad -11.56$$

Since the answer cannot be negative,

$$x = 9.56 \text{ cm}$$

and

$$x + 2 = 11.56 \text{ cm}$$

∴ the rectangle has adjacent sides equal to 9.56 cm and 11.56 cm.

EXAMPLE 16

The template shown in Fig. 2.2 has an area of 60 cm². Find the radius of the template. (Take $\pi = \frac{22}{7}$.)

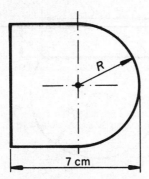

Fig. 2.2

The template consists of a semi-circle whose area is $\dfrac{\pi R^2}{2}$ and a rectangle whose area is $2R(7 - R)$.

$$\therefore \qquad \frac{\pi R^2}{2} + 2R(7 - R) = 60$$

$$\pi R^2 + 4R(7 - R) = 120$$

$$\frac{22R^2}{7} + 28R - 4R^2 = 120$$

$$22R^2 + 196R - 28R^2 = 840$$

$$-6R^2 + 196R - 840 = 0$$

$$6R^2 - 196R + 840 = 0$$

or $\qquad\qquad 3R^2 - 98R + 420 = 0$

Here $a = 3$, $b = -98$ and $c = 420$,

$$\therefore \qquad R = \frac{-(-98) \pm \sqrt{(-98)^2 - 4 \times 3 \times 420}}{2 \times 3}$$

$$\therefore \qquad R = 5.1 \quad \text{or} \quad 27.6$$

From Fig. 2.2 it will be seen that R must be less than 7 cm.

Hence: $\qquad R = 5.1$ cm

EXAMPLE 17

A section of an air duct is shown by the full lines in Fig. 2.3.

(a) Show that:

$$w^2 - 2Rw + \frac{R^2}{4} = 0.$$

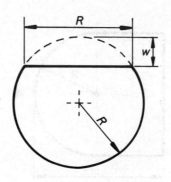

Fig. 2.3

(b) Find the value of w when $R = 2$ m.

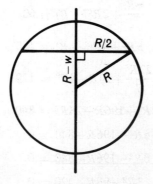

Fig. 2.4

Using the construction shown in Fig. 2.4 we have, by Pythagoras,

$$(R-w)^2+\left(\frac{R}{2}\right)^2 = R^2$$

$$R^2-2Rw+w^2+\frac{R^2}{4} = R^2$$

$$\therefore \qquad w^2-2Rw+\frac{R^2}{4} = 0$$

When $R = 2$,

$$w^2-4w+1 = 0$$

Here $a = 1$, $b = -4$ and $c = 1$

$$\therefore \qquad w = \frac{-(-4)\pm\sqrt{(-4)^2-4\times1\times1}}{2\times1}$$

Hence $\qquad w = 3.732 \quad \text{or} \quad 0.268 \text{ m.}$

Now w must be less than 2 m

$$\therefore \qquad w = 0.268 \text{ m}$$

Exercise 4

1) The length L of a wire stretched tightly between two supports in the same horizontal line is given by:

$$L = S+\frac{8D^2}{3S}$$

where S is the span and D is the (small) sag. If $L = 150$ and $D = 5$ find the value of S.

2) In a right-angled triangle the hypotenuse is twice as long as one of the sides forming the right angle. The remaining side is 8 cm long. Find the length of the hypotenuse.

3) The area of a rectangle is 61.75 m². If the length is 3 m more than the width find the dimensions of the rectangle.

4) The total surface area of a cylinder whose height is 7.5 cm is 290 cm². Find the radius of the cylinder.

5) If a segment of a circle has a radius R, a height H and a length of chord W show that

$$R = \frac{W^2}{8H}+\frac{H}{2}$$

Rearrange this equation to give a quadratic equation for H and hence find H when $R = 12$ cm and $W = 8$ cm.

6)   Fig. 2.5 shows a template whose area is 10 690 mm². Find the value of r taking $\pi = \frac{22}{7}$.

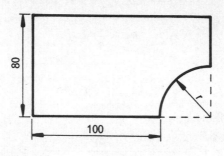

Fig. 2.5

7)   The total iron loss in a transformer is given by the equation $P = 0.1f + 0.006f^2$. If $P = 20$ watts find the value of the frequency f.

8)   The volume of a frustum of a cone is given by the formula $V = \frac{1}{3}\pi h(R^2 + rR + r^2)$ where h is the height of the frustum and R and r are the radii at the large and small ends respectively. If $h = 9$ cm, $R = 4$ cm and the volume is 337.2 cm³ what is the value of r?

9)   A pressure vessel is of the shape shown in Fig. 2.6, the radius of the vessel being r mm. If the surface area is 30 000 mm³ find r.

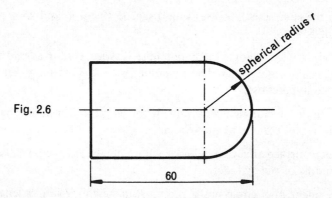

Fig. 2.6

10)   A square steel plate is pierced by a tool leaving a margin of 2 cm all round. The area of the hole is one third that of the original plate. What are the dimensions of the original plate?

FLOW CHARTS AND ELECTRONIC CALCU-LATING MACHINES

3.

After reaching the end of this chapter you should be able to :-

1. *Draw a flow chart to show a sequence of arithmetic operations.*
2. *Use the standard symbols for flow charts.*
3. *Perform chain calculations.*
4. *Rearrange a problem to simplify the calculation.*

FLOW CHARTS

There are many times when a series of operations or instructions has to be carried out. If these are written down in sequence then this list is called a *programme*.

A simple and clear way of giving a programme is by using a *flow chart*. An example of a flow chart is shown in Fig. 3.1.

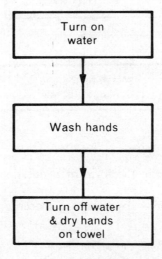

Fig. 3.1

The instructions are given in rectangles which are connected by arrows showing the order in which to proceed.

It is usual to use a rectangle having rounded ends for the first and last instructions. An example of this is shown in Fig. 3.2.

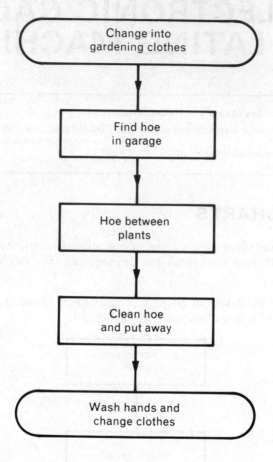

Fig. 3.2

An instruction may be in the form of a question having two possible answers — yes or no.

A question is shown on a flow chart by a decision symbol as shown in Fig. 3.3.

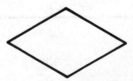

Fig. 3.3

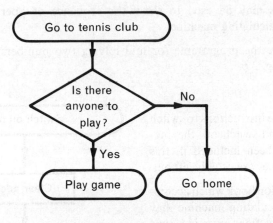

Fig. 3.4

The flow chart in Fig. 3.4 shows a decision symbol leading to two separate results.

Sometimes the decision symbol may result in some of the instructions being repeated.

An example of this is shown in Fig. 3.5. In this case the two middle instructions may be repeated and they form what is called a *loop*.

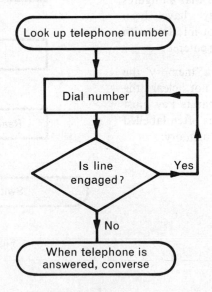

Fig. 3.5

Flow charts may be used to show the sequence of operations on an electronic calculating machine.

For example the programme for multiplying two numbers, i.e., $a \times b$ is shown in Fig. 3.6.

Although the instructions to switch on, clear, and switch off the machine have been included in this flow chart they are usually left out.

The instruction book which accompanies a calculating machine may include instructions given in a type of flow chart. This is often a 'step by step' set of instructions giving the sequence in which the keys should be pressed.

Instructions for the multiplication of 8.24 by 76.91 may look like this:

The first instruction is the 'clear key' which ensures that all figures entered previously have been erased, and will not interfere with the new data being entered.

If a machine has a 'memory' the C key may not clear the memory and a separate key must be pressed — this is often labelled CM, i.e., clear memory.

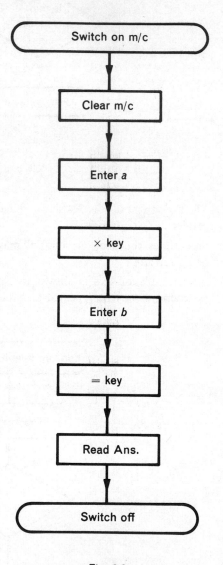

Fig. 3.6

CHAIN CALCULATIONS

The result of $5 \times 4 \div 2$ could be obtained by the sequence shown in Fig. 3.7. Most calculators, however, do not require the use of the $=$ key when followed by an operation key, i.e. $+$, $-$, $\times$, or $\div$ so that the fourth operation may be omitted. The result of 5×4 is displayed as soon as the $\div$ key is pressed.

This type of calculation is called a 'chain calculation' and the modified sequence is shown in Fig. 3.8.

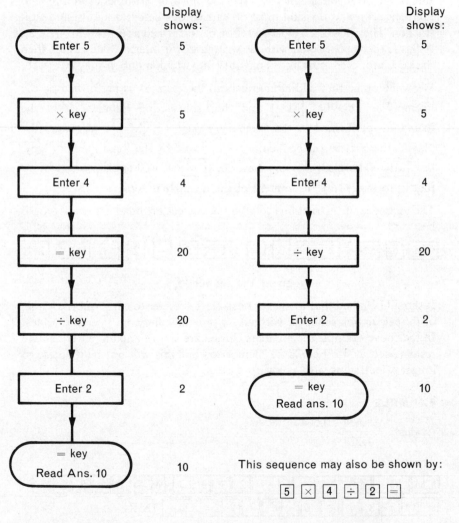

Fig. 3.7   Fig. 3.8

Care must be taken when carrying out chain calculations to make sure that the sequence of operations used will give the correct arithmetical result.

For example consider the value of $6 + 4 \times 3 - 5 \div 2$.

If the sequence used is:

$$\boxed{6}\ \boxed{+}\ \boxed{4}\ \boxed{\times}\ \boxed{3}\ \boxed{-}\ \boxed{5}\ \boxed{\div}\ \boxed{2}\ \boxed{=}$$

then the machine will perform each operation in turn, that is to add 4 to 6 giving 10, multiply by 3 giving 30, subtract 5 giving 25 and then divide by 2 giving a final result of 12.5.

But in an arrangement of this type the rules of arithmetic require that multiplication and division must be carried out before addition and subtraction. The calculation must therefore be performed as $6 + (4 \times 3) - (5 \div 2)$. If this is to be completed without writing down intermediate results then the calculator used must have a memory into which numbers may be added.

We will assume that adding numbers to the memory is performed by the 'memory+' key, i.e. $\boxed{\text{M+}}$ and that subtracting numbers from the memory is performed by the 'memory−' key, i.e. $\boxed{\text{M−}}$; also that displaying the content of the memory is achieved by the 'read memory' key, i.e. $\boxed{\text{MR}}$. The 'clear memory' key, i.e. $\boxed{\text{CM}}$ is used to clear the memory that is, to make sure the content of the memory is zero.

The sequence of instructions to give the correct arithmetical result would be:

$$\boxed{\text{CM}}\ \boxed{6}\ \boxed{\text{M+}}\ \boxed{4}\ \boxed{\times}\ \boxed{3}\ \boxed{=}\ \boxed{\text{M+}}\ \boxed{5}\ \boxed{\div}\ \boxed{2}\ \boxed{=}\ \boxed{\text{M−}}\ \boxed{\text{MR}}$$

giving the answer 15.5

It should be noted that whenever possible it is better to carry out the whole of the calculation sequence without writing down any intermediate values. In the above example the numbers chosen are simple and any intermediate results such as 4×3 could be memorised but this will not be the case in longer calculations with awkward figures.

EXAMPLE 1

Evaluate $\dfrac{0.674}{1.239} - \dfrac{0.564 \times 1.89}{0.379}$

The sequence of operations is:

giving an answer $-2.268\ 57$.

REARRANGING A PROBLEM TO EASE CALCULATIONS

It is sometimes necessary to change the order in which a problem is given so that it may be performed completely on the calculating machine.

EXAMPLE 2

Evaluate $9.7 + \dfrac{55.15}{29.6 - 8.64}$.

If the 9.7 is entered first into memory, it is not possible to perform the division. However there is no difficulty if the problem is rewritten as $\dfrac{55.15}{29.6 - 8.64} + 9.7$ and the sequence of operation is:

$$\boxed{\text{CM}}\ \boxed{2}\ \boxed{9}\ \boxed{\cdot}\ \boxed{6}\ \boxed{-}\ \boxed{8}\ \boxed{\cdot}\ \boxed{6}\ \boxed{4}\ \boxed{=}\ \boxed{\text{M}+}\ \boxed{5}\ \boxed{5}\ \boxed{\cdot}\ \boxed{1}\ \boxed{5}\ \boxed{\div}$$

$$\boxed{\text{MR}}\ \boxed{=}\ \boxed{+}\ \boxed{9}\ \boxed{\cdot}\ \boxed{7}\ \boxed{=}$$

giving an answer 12.331 2.

It will be noticed that the bottom line of the fraction is worked out first, then put into memory — then after entering the 55.15 the content of the memory is recalled and divided into the 55.15 etc.

It is often necessary to adopt this method of dealing with the denominator before tackling the numerator.

THE RECIPROCAL KEY

The reciprocal of x is $\dfrac{1}{x}$, and most calculators have a 'reciprocal key' denoted by $\boxed{\frac{1}{x}}$. The use of this key may avoid the necessity of changing the order of the problem.

EXAMPLE 3

Evaluate $\dfrac{5.79 - 0.673}{2.022 + 8.63}$

As in Example 2 the solution is straightforward providing the denominator is considered first.

It is possible, however, to proceed by starting work on the top line, putting the result into memory; then evaluating the denominator and dividing the result by the content of the memory. This will give the value of $\dfrac{2.022+8.63}{5.79-0.673}$ which is the reciprocal of the given problem. Use of the reciprocal $\boxed{\frac{1}{x}}$ key will obtain the required result. The full sequence would be:

$\boxed{\text{CM}}\boxed{5}\boxed{\cdot}\boxed{7}\boxed{9}\boxed{-}\boxed{0}\boxed{\cdot}\boxed{6}\boxed{7}\boxed{3}\boxed{=}\boxed{\text{M+}}\boxed{2}\boxed{\cdot}\boxed{0}\boxed{2}\boxed{2}\boxed{+}$
$\boxed{8}\boxed{\cdot}\boxed{6}\boxed{3}\boxed{=}\boxed{\div}\boxed{\text{MR}}\boxed{=}\boxed{\frac{1}{x}}$ giving an answer 0.480 38

A flow chart containing a loop is required for the next example.

EXAMPLE 4

Find the total mass of:

> 5 castings, each of 30 kg mass
>
> 4 castings, each of 20 kg mass
>
> 3 castings, each of 15 kg mass
>
> 6 castings, each of 10 kg mass

The total mass will be given by $5\times30+4\times20+3\times15+6\times10$. We have added the products of the number of castings and their respective masses.

In mathematical notation we show this as $\Sigma(n\times m)$ where n is the number of castings having a particular mass m. Σ is the capital Greek letter 'sigma' and means 'the sum of'.

The flow chart, Fig. 3.9, shows how to use the calculating machine to perform the operation $\Sigma(n\times m)$, the values of n and m being entered in pairs.

It will be necessary for the calculating machine used in this type of question to have a memory into which numbers may be added.

In the first run through the flow chart the result of 5×30, i.e. 150, will be added to zero in the memory giving a total of 150. The second time the result of 4×20, i.e. 80, will be added to the 150 already in the memory giving a new total of 230.

This sequence continues until the given list is completed, and at the end the memory will contain the answer, i.e. 335 kg.

The sequence of operations would be:

$\boxed{\text{C}}\boxed{\text{CM}}\boxed{5}\boxed{\times}\boxed{3}\boxed{0}\boxed{=}\boxed{\text{M+}}\boxed{4}\boxed{\times}\boxed{2}\boxed{0}\boxed{=}\boxed{\text{M+}}\boxed{3}\boxed{\times}$
$\boxed{1}\boxed{5}\boxed{=}\boxed{\text{M+}}\boxed{6}\boxed{\times}\boxed{1}\boxed{0}\boxed{=}\boxed{\text{M+}}\boxed{\text{MR}}$

giving an answer 335.

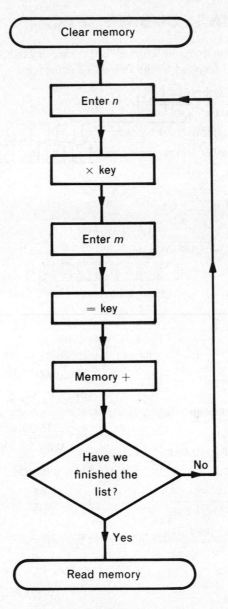

Fig. 3.9

SQUARE ROOTS

Some calculating machines have a square root key denoted by $\boxed{\sqrt{}}$. Although this key is not absolutely necessary if the machine has logarithmic facilites it is extremely useful being straightforward and quick to use.

SQUARING, CUBING, ETC.

This operation may be carried out using logarithmic keys but many machines will respond to the following sequences:

Squaring a number $\boxed{\times}$ $\boxed{=}$

Cubing a number $\boxed{\times}$ $\boxed{=}$ $\boxed{=}$

Raising a number to the 4th power $\boxed{\times}$ $\boxed{=}$ $\boxed{=}$ $\boxed{=}$ or $\boxed{\times}$ $\boxed{=}$ $\boxed{\times}$ $\boxed{=}$

Raising a number to the 5th power $\boxed{\times}$ $\boxed{=}$ $\boxed{=}$ $\boxed{=}$ $\boxed{=}$

and so on . . .

EXAMPLE 5

Find the value of 2.71^3.

The sequence of operations is:

$$\boxed{2} \boxed{\cdot} \boxed{7} \boxed{1} \boxed{\times} \boxed{=} \boxed{=}$$

giving an answer 19.902 51.

Exercise 5

Evaluate:

1) $\dfrac{0.378\,6 \times 0.039\,72}{31.67}$

2) $\dfrac{97.61 \times 0.000\,46}{0.091\,74}$

3) $\dfrac{0.86 \times 298.7 \times 0.683 \times 15}{39.4}$

4) $\dfrac{0.014\,6 \times 0.798 \times 643}{33\,000 \times 11.8}$

5) $\dfrac{9.87 \times 30 \times 10^6 \times 0.048}{70^2}$

6) $(0.036\,14)^2$

7) $(0.785)^3$

8) $(0.001\,53)^4$

9) $\sqrt{0.256\,9}$

10) $\dfrac{2.81}{0.64} - \dfrac{7.665}{1.006}$

11) $\sqrt{8.495^2 - 3.431^2}$

12) $\dfrac{18.84}{0.006} - (456.7)(5.2)^2$

13) $0.678\,9 + \sqrt{\dfrac{6.32^3}{1.7}}$

14) $\dfrac{64.57 + (0.987)(346.7)}{(6.39)(43.21) - 8.251}$

15) Find the total cost of:

   10 articles at £5.64 each,

   7 articles at £3.47 each,

   23 articles at £0.92 each, and

   17 articles at £2.43 each.

16)   Find the average mass of:

5 castings, each of mass 67 kg,

9 castings, each of mass 22 kg,

3 castings, each of mass 107 kg,

7 castings, each of mass 76 kg, and

4 castings, each of mass 59 kg.

17)   The formula for finding the distance $\bar{x}$ of the centroid of a cross-sectional area from the y-axis is

$$\frac{\Sigma\, Ax}{\Sigma\, A}.$$

Find $\bar{x}$ if the corresponding values of A and x are:

A (cm²)	x (cm)
25.2	6.9
43.7	2.5
9.8	3.6
12.6	1.3

SUMMARY

a) Flow chart symbols:

First and last instructions

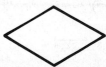

Intermediate instructions

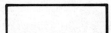

Decisions

b) In all mathematics the symbol Σ means 'the sum of'.

LOGARITHMS

After reaching the end of this chapter you should be able to:-

1. *State the laws of indices:*

   $a^m \times a^n = a^{m+n}, \dfrac{a^m}{a^n} = a^{m-n}, (a^m)^n = a^{mn},$

   $\sqrt[m]{a^n} = a^{m/n}, \dfrac{1}{a^n} = a^{-n}, a^0 = 1.$

2. *Solve equations containing logarithms.*
3. *State the rules for the use of logarithms.*
4. *Use logarithms to the base e.*
5. *Solve indical equations.*

LAWS OF INDICES

Multiplication: $a^m \times a^n = a^{m+n}$.

Division: $\dfrac{a^m}{a^n} = a^{m-n}$.

Powers: $(a^m)^n = a^{mn}$.

Roots: $\sqrt[m]{a^n} = a^{n/m}$.

Reciprocals: $\dfrac{1}{x^n} = x^{-n}$.

Zero index: $a^0 = 1$.

EXAMPLE 1

a) $a^3 \times a^4 \times a^7 = a^{3+4+7} = a^{14}$.

b) $\dfrac{b^4 c^6}{b^2 c^9} = \dfrac{b^{4-2}}{c^{9-6}} = \dfrac{b^2}{c^3}$.

c) $\sqrt[3]{27a^6} = (3^3 a^6)^{1/3} = 3^{3 \times 1/3} a^{6 \times 1/3} = 3a^2$.

d) $(2x^2)^{-3} = \dfrac{1}{(2x^2)^3} = \dfrac{1}{2^3 x^6}$

e) $37^0 = 1$

Exercise 6

Simplify each of the following with positive indices:

1) $a^5 \times a^6$

2) $z^4 \times z^7$

3) $y^3 \times y^4 \times y^5$

36

4) $2^3 \times 2^5$

5) $3 \times 3^2 \times 3^5$

6) $\frac{1}{2}a \times \frac{1}{4}a^2 \times \frac{3}{4}a^3$

7) $a^5 \div a^2$

8) $m^{12} \div m^5$

9) $2^8 \div 2^4$

10) $x^{20} \div x^5$

11) $a^5 \times a^3 \div a^4$

12) $q^7 \times q^6 \div q^5$

13) $\dfrac{m^5}{m^3} \times \dfrac{m}{m^2}$

14) $\dfrac{l^5 \times l^6}{l^2 \times l^7}$

15) $\dfrac{aL^4}{aL^2}$

16) $(x^3)^4$

17) $(a^5)^3$

18) $(3x^4)^2$

19) $(2^3)^2$

20) $(10^3)^2$

21) $(ab^2)^3$

22) $(ab^2c^3)^4$

23) $(2x^2y^3z)^5$

24) $\left(\dfrac{3m^2}{4n^3}\right)^5$

25) Find the values of 10^{-1}, 2^{-2}, 3^{-4}, 5^{-2}.

26) Find the values of $2^4 \times 2$, $5^2 \times 5^{1/2} \times 5^{3/2}$, $8^{1/3}$, $27^{1/3}$.

27) Express as powers of 3: 9^3, 27^5, 81^3, $9^4 \times 27^3$.

28) Express as powers of a: $\sqrt[5]{a}$, $\sqrt[3]{a^2}$, $\sqrt[7]{a^4}$, $\sqrt{a^6}$.

29) Find the values of: $2^4 \times 2^2$, $(10^2)^3$, $(2^5)^{2/5}$, $64^{1/2}$.

30) Evaluate: $32^{2/5}$, $16^{3/4}$, $(81)^{-1/4}$, $9^{-1/2}$.

31) Find the value of $32^{3/5} \times 25^{1/2} \times 64^{-1/3}$.

32) Find the value of $4^{-1/2} + \left(\dfrac{1}{27}\right)^{1/3}$.

33) Find the values of $\left(\dfrac{1}{5}\right)^0$, $(125)^{-1/3}$, $(1\,000\,000)^{5/6}$.

34) Simplify: $(9x^4)^{-1/2}$, $(27x^6)^{-1/3}$, $(25a^8)^{-1/2}$.

35) Simplify: $(64a^6)^{1/2} \div (64a^6)^{-1/3}$.

THEORY OF LOGARITHMS

If N is a number such that

$$N = a^x$$

we say that x is the logarithm of N to the base a. We write

$$\log_a N = x$$

It should be carefully noted that

$$\text{number} = \text{base}^{\text{logarthim}}$$

Since $125 = 5^3$ we may write $\log_5 125 = 3$ and because $16 = 2^4$ we may write $\log_4 16 = 2$.

EXAMPLE 2

a)   If $\log_7 49 = x$, find the value x.

Writing the equation in index form we have,

$$49 = 7^x$$
$$7^2 = 7^x$$

Hence $x = 2$ (since the indices on each side of the equation must be the same).

b)   If $\log_x 8 = 3$, find the value of x.

Writing the equation in index form we have,

$$8 = x^3$$
$$2^3 = x^3$$

Hence $x = 2$ (since the indices on both sides of the equation are the same the bases must be the same).

Exercise 7

In each of the following find the value of x:

1) $\log_x 9 = 2$ 5) $\log_3 x = 2$ 9) $\log_x 8 = 3$

2) $\log_x 81 = 4$ 6) $\log_4 x = 3$ 10) $\log_x 27 = 3$

3) $\log_2 16 = x$ 7) $\log_{10} x = 2$

4) $\log_5 125 = x$ 8) $\log_7 x = 0$

RULES FOR THE USE OF LOGARITHMS

These rules are true for any chosen value of the base:

1)   The logarithm of two numbers multiplied together may be found by adding their individual logarithms:

$$\log xy = \log x + \log y$$

2) The logarithm of two numbers divided may be found by subtracting their individual logarithms:

$$\log \frac{x}{y} = \log x - \log y$$

3) The logarithm of a number raised to a power may be found by multiplying the power by the logarithm of the number:

$$\log x^n = n \cdot \log x$$

LOGARITHMS TO THE BASE 10

Logarithms to the base 10 are called common logarithms and are given as $\log_{10}$. When using logarithmic tables to solve numerical problems, the values given in the tables will be to the base 10. It is assumed that the reader is familiar with the use of common logarithms.

LOGARITHMS TO THE BASE 'e'

In higher mathematics all logarithms are taken to the base e, where e = 2.718 28. Logarithms to this base are often called natural logarithms. They are also called Naperian or hyperbolic logarithms.

To avoid confusion the base of a logarithm should always be stated, e.g. $\log_{10}$, but in practice this is often omitted if it is reasonably obvious which base is being used.

In the examples which follow the base will always be given.

Common logarithms are given as $\log_{10}$ (or lg)

and natural logarithms are given as $\log_e$ (or ln).

CHOICE OF BASE

When using logarithmic tables for numerical calculations common logarithms, i.e. $\log_{10}$, are preferred. This is because they are simpler to use in table form than natural logarithms.

If an electronic calculating machine of the scientific type (i.e., having keys which give trignometrical and logarithmic functions in addition to the usual +, −, × and ÷ etc.) is used then it is just as easy to use logarithms to the base e. Some machines have keys for both $\log_e$ and $\log_{10}$ but on the more limited models only $\log_e$ is given.

The natural logarithms is found by using the $\log_e$ (or ln) key and the natural antilogarithm is found by using the e^x key.

In the worked examples which follow in this chapter two alternative methods of solution are given — one using logarithmic tables and common logarithms, and the other using calculating machine and natural logarithms.

EXAMPLE 3

Evaluate $3.714^{2.87}$.

Let $x = 3.714^{2.87}$

and taking logarithms of both sides we have:

$$\log x = \log 3.714^{2.87}$$

$\therefore$ $\log x = 2.87 \times \log 3.714$

$\therefore$ $x = \text{antilog}\,(2.87 \times \log 3.714)$

The base of the logarithms has not yet been chosen — the sequence given will be true for any base value.

(i) Calculations using common logarithmic tables, i.e. $\log_{10}$:

$x = \text{antilog}_{10}(2.87 \times \log_{10} 3.714)$

$\ = \text{antilog}_{10}(2.87 \times 569\,9)$

$\ = \text{antilog}_{10}\,1.636$

$\ = 43.25$

Number	$\log_{10}$
2.87	0.457 9
0.569 9	$\bar{1}.755\,8$
1.636	0.213 7

or alternatively:

(ii) Calculations using a machine with natural logarithms, i.e. $\log_e$, the sequence of operations being given on a flow chart (see opposite):

It should be noted that the answers obtained by the two methods are slightly different. This is because the tables used are only of the four figure type, and accuracy of the last figure cannot be guaranteed.

The sequence of operations would then be:

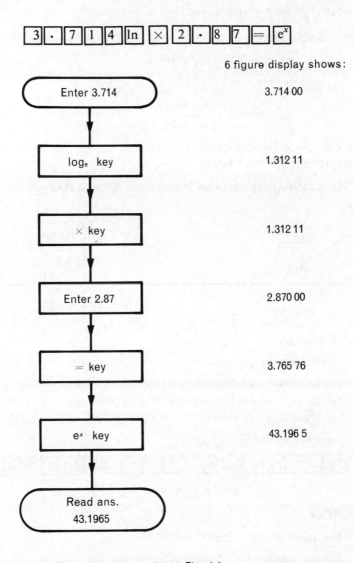

Fig. 4.1

INDICIAL EQUATIONS

These are equations in which the number to be found is an index, or part of an index.

The method of solution is to reduce the given equation to an equation involving logarithms, as the following examples will illustrate:

EXAMPLE 4

If $8.79^x = 67.65$ find the value of x.

Now taking logarithms of both sides of the given equation we have

$$\log 8.79^x = \log 67.65$$

$\therefore$ $\qquad\qquad x \,.\, \log 8.79 = \log 67.65$

$\therefore$ $\qquad\qquad x = \dfrac{\log 67.65}{\log 8.79}$

The base of the logarithms has not yet been chosen, the above procedure being true for any base value.

(i) Calculations using common logarithmic tables, i.e. $\log_{10}$:

$$x = \frac{\log_{10} 67.65}{\log_{10} 8.79}$$

$$= \frac{1.830\,2}{0.944\,0}$$

$$= 1.938$$

Number	$\log_{10}$
1.830 2	0.262 5
0.944 0	$\overline{1}.975\,0$
1.938	0.287 5

or alternatively:

(ii) Calculations using a machine with natural logarithms, i.e. $\log_e$:

The quickest way, that is without using the reciprocal $\boxed{\frac{1}{x}}$ key, is to find the value of the bottom line and add this into the memory. Then find the value of the top line and divide this by the content of the memory.

This sequence would be:

$$\boxed{\text{CM}}\;\boxed{8}\;\boxed{\cdot}\;\boxed{7}\;\boxed{9}\;\boxed{\ln}\;\boxed{\text{M+}}\;\boxed{6}\;\boxed{7}\;\boxed{\cdot}\;\boxed{6}\;\boxed{5}\;\boxed{\ln}\;\boxed{\div}\;\boxed{\text{MR}}\;\boxed{=}$$

giving an answer 1.938 86

EXAMPLE 5

Find the value of x if $1.793^{x+3} = 20^{0.982}$

Now taking logarithms of both sides of the given equation we have:

$$\log 1.793^{x+3} = \log 20^{0.982}$$

$\therefore$ $\qquad (x+3)(\log 1.793) = (0.982)(\log 20)$

$\therefore$ $\qquad\qquad x+3 = \dfrac{(0.982)(\log 20)}{\log 1.793}$

$\therefore$ $\qquad\qquad x = \dfrac{(0.982)(\log 20)}{\log 1.793} - 3$

The base of the logarithms has not yet been chosen, the above procedure being true for any base value.

(i) Calculations using common logarithic tables, i.e. $\log_{10}$:

$$x = \frac{(0.982)(\log_{10} 20)}{\log_{10} 1.793} - 3$$

$$= \frac{(0.982)(1.301\,0)}{0.253\,6} - 3$$

$$= 5.032 - 3$$

$$= 2.032$$

Number	$\log_{10}$
0.982	$\bar{1}.991\,7$
1.301 0	0.114 2
	0.105 9
0.253 6	$\bar{1}.404\,1$
5.032	0.701 8

or alternatively;

(ii) Calculations using a machine with natural logarithms, i.e. $\log_e$:

The procedure will be similar to that used in Example 4. The sequence of operations is then:

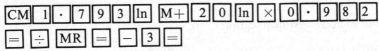

giving an aswer 2.038 29

Exercise 8

Evaluate the following:

1) $\sqrt[3]{9.253}$

2) $\sqrt[4]{0.023\,8}$

3) $\sqrt{0.817\,6^3}$

4) $\sqrt[5]{7.891^3}$

5) $0.062\,9^{2/5}$

6) $0.371\,6^{7/3}$

7) $\sqrt{\dfrac{11.16 \times 0.329\,8}{7.315 \times 0.897\,4}}$

8) $\sqrt[4]{\dfrac{0.352 \times 17.26^2}{11.15 \times 0.93^3}}$

9) $[(0.816\,3)^2 \times (7.315)^4]^{1/5}$

10) $\dfrac{0.315\,7^2 \times \sqrt{18.16}}{\sqrt[3]{0.001\,76} \times 49.18}$

11) $11.57^{0.3}$

12) $15.62^{2.15}$

13) $0.632\,7^{0.5}$

14) $0.065\,21^{3.16}$

15) $27.15^{-0.4}$

16) $73.2^{0.298}$

17) $0.598\ 7^{-2.18}$

18) $3.761^{-1.79}$

19) $0.211^{-0.816}$

20) $\dfrac{5}{0.316\ 8^{0.25}}$

21) $\dfrac{0.798^{0.3}}{18.16^{0.2}}$

22) $\dfrac{37.5}{0.25^{0.2} \times 0.03^{0.4}}$

23) $28.7^{0.62} \times 0.632$

24) $\dfrac{3.261^{0.3}}{\sqrt[6]{2.817}}$

Find the value of x in the following:

25) $3.6^x = 9.7$

26) $0.9^x = 2.176$

27) $\left(\dfrac{1}{7.2}\right)^x = 1.89$

28) $1.4^{(x+2)} = 9.3$

29) $21\ 9^{(3-x)} = 7.334$

30) $2.79^{(x-1)} = 4.377^x$

31) $\left(\dfrac{1}{0.64}\right)^{(2+x)} = 1.543^{(x+1)}$

32) $\dfrac{1}{0.9^{(x-2)}} = 8.45$

SUMMARY

a) Law of indices: Multiplication $a^m \times a^n = a^{m+n}$

Division $\dfrac{a^m}{a^n} = a^{m-n}$

Powers $(a^m)^n = a^{mn}$

Roots $\sqrt[m]{a^n} = a^{n/m}$

Reciprocals $\dfrac{1}{a^n} = a^{-n}$

Zero index $a^0 = 1$

b) Logarithms: If $N = a^x$

then $\log_a N = x$

c) Remember that   number $=$ base ^{logarithm}

d) Rules for use of logarithms:

$$\log xy = \log x + \log y$$

$$\log \frac{x}{y} = \log x - \log y$$

$$\log x^n = n \cdot \log x$$

e) Common logarithms are denoted by: $\log_{10}$ (or lg).

f) Natural logarithms are denoted by: $\log_e$ (or ln), where e $= 2.718\,28$.

Self-test 1

1) $\dfrac{a^4 \times a^3}{a^2}$ is equal to:

    a a^6 **b** a^{10} **c** $a^{7/2}$ **d** a^5

2) $^2\sqrt{a^6}$ is equal to:

    a a^3 **b** a^4 **c** a^8 **d** a^{12}

3) $\sqrt{\dfrac{1}{x^6}}$ is equal to:

    a $\dfrac{1}{x^4}$ **b** $\dfrac{1}{x^3}$ **c** x^{-3} **d** x^{-4}

4) $\left(\dfrac{1}{3}\right)^{-2}$ is equal to:

    a $\dfrac{1}{9}$ **b** $\dfrac{1}{6}$ **c** $-\dfrac{1}{6}$ **d** 9

5) $\left(\dfrac{1}{16}\right)^{-1/2}$ is equal to:

    a 4   **b** 8   **c** $-\dfrac{1}{32}$ **d** $-\dfrac{1}{4}$

6) $x^0 \times \dfrac{1}{x}$ is equal to:

    a 1   **b** 0   **c** $\dfrac{1}{x}$ **d** x^{-1}

7) $\log_2 8$ is equal to:

    a 2   **b** 3   **c** 4   **d** 16

8) $\log_3 \dfrac{1}{9}$ is equal to:

    a $\dfrac{1}{27}$ **b** $\dfrac{1}{3}$ **c** $\dfrac{1}{2}$ **d** -2

9) If $\log_b 27 = 3$ then the value of b is:

    a 3   **b** 9   **c** $\dfrac{1}{9}$ **d** -3

10) If $\log_2 x = 5$ then the value of x is:

    a 32   **b** 25   **c** 10   **d** 2.5

11) $\log \dfrac{3 \times 4}{2}$ is equal to:

    a $\log 6$ **b** $\log (3 \times 4 - 2)$ **c** $\dfrac{\log 3 \times \log 4}{\log 2}$ **d** $\dfrac{\log 12}{\log 2}$

12) $\log 6 - \log 3$ is equal to:

    a $\log 3$ **b** $\dfrac{\log 6}{\log 3}$ **c** $\log 2$ **d** $\log 6 + \log (-3)$

13) $\log_2 1$ is equal to:

    a 2 **b** 1 **c** $\dfrac{1}{2}$ **d** 0

14) $\log_2 2$ is equal to:

    a 4 **b** 2 **c** 1 **d** 0

5. EXPONENTIAL RELATIONSHIPS

After reaching the end of this chapter you should be able to :-
1. *Determine the Napierian logarithm for any number.*
2. *Deduce that $\log_e N = 2.3026 \times \log_{10} N$.*
3. *Determine the value of a positive number given its Napierian logarithm.*
4. *Calculate the numerical value of Ae^{bx} where A is positive and b can be either positive or negative.*

NATURAL LOGARITHMIC TABLES

In most books of mathematical tables there is a table of natural logarithms. Part of such a table is shown below:

Hyperbolic, Natural or Naperian Logarithms:

	0	1	2	3	4	5	6	7	8	9									
4.5	1.5041	5063	5085	5107	5129	5151	5173	5195	5217	5239	2	4	7	9	11	13	15	18	20
4.6	1.5261	5282	5304	5326	5347	5369	5390	5412	5433	5454	2	4	6	9	11	13	15	17	19
4.7	1.5476	5497	5518	5539	5560	5581	5602	5623	5644	5665	2	4	6	8	11	13	15	17	19
4.8	1.5686	5707	5728	5748	5769	5790	5810	5831	5851	5872	2	4	6	8	10	12	14	16	19
4.9	1.5892	5913	5933	5953	5974	5994	6014	6034	6054	6074	2	4	6	8	10	12	14	16	18
5.0	1.6094	6114	6134	6154	6174	6194	6214	6233	6253	6273	2	4	6	8	10	12	14	16	18
5.1	1.6292	6312	6332	6351	6371	6390	6409	6429	6448	6467	2	4	6	8	10	12	14	16	18
5.2	1.6487	6506	6525	6544	6563	6582	6601	6620	6639	6658	2	4	6	8	10	11	13	15	17
5.3	1.6677	6696	6714	6734	6752	6771	6790	6808	6827	6845	2	4	6	7	9	11	13	15	17
5.4	1.6864	6882	6901	6919	6938	6956	6974	6993	7011	7029	2	4	5	7	9	11	13	15	17

Natural logarithms of 10^{+n}:

n	1	2	3	4	5	6	7	8	9
$\log_e 10^n$	2.3026	4.6052	6.9078	9.2103	11.5129	13.8155	16.1181	18.4207	20.7233

Natural logarithms of 10^{-n}:

n	1	2	3	4	5	6	7	8	9
$\log_e 10^{-n}$	$\bar{3}.6974$	$\bar{5}.3948$	$\bar{7}.0922$	$\bar{10}.7897$	$\bar{12}.4871$	$\bar{14}.1845$	$\bar{17}.8819$	$\bar{19}.5793$	$\bar{21}.2767$

The first column of a set of full tables gives the natural logarithms of numbers from 1.0 to 9.9 (the specimen table above only gives numbers from 4.5 to 5.49), and the tables are read in the same way as ordinary log tables except that the characteristic is also given. Thus:

$$\log_e 4.568 = 1.5191$$

When a natural logarithm of a number, which lies outside the tabulated range, is required the subsidiary table has to be used. The following examples show how this is done.

EXAMPLE 1

To find $\log_e 483.4$.

$$483.4 = 4.834 \times 100 = 4.834 \times 10^2$$

$\therefore$ $\log_e 483.4 = \log_e 4.834 + \log_e 10^2$

From the main table: $\log_e 4.834 = 1.575\,6$.

From the subsidiary table: $\log_e 10^2 = 4.605\,2$.

$\therefore$ $\log_e 483.4 = 1.575\,6 + 4.605\,2 = 6.180\,8$

EXAMPLE 2

To find $\log_e 0.053\,61$.

$$0.053\,61 = \frac{5.361}{100} = \frac{5.361}{10^2} = 5.361 \times 10^{-2}$$

$\therefore$ $\log_e 0.053\,61 = \log_e 5.361 + \log_e 10^{-2}$

From the main table: $\log_e 5.361 = 1.679\,2$

From the subsidiary table: $\log_e 10^{-2} = \bar{5}.394\,8$

$\therefore$ $\log_e 0.053\,61 = 1.679\,2 + \bar{5}.394\,8 = \bar{3}.074\,0$

$$= -3 + 0.074\,0 = -2.926\,0$$

CONVERSION OF NATURAL LOGARITHMS TO COMMON LOGARITHMS

The use of tables of natural logarithms is rather tedious as the last two worked examples show. Their use may be avoided by finding the common logarithm and then converting to a natural logarithmic value.

Suppose we have a number, N, and wish to find the value of its natural logarithm, that is $\log_e N$.

Let $x = \log_e N$

then: $e^x = N$

and taking common logarithms of both sides we have

$$\log_{10} e^x = \log_{10} N$$

$\therefore$ $x \cdot \log_{10} e = \log_{10} N$

$\therefore$ $x = \dfrac{\log_{10} N}{\log_{10} e}$

or: $\log_e N = \dfrac{\log_{10} N}{\log_{10} e}$

The value of $\log_{10} e = \log_{10} 2.718\,3 = 0.434\,3$

---

hence: $$\log_e N = \frac{\log_{10} N}{0.434\,3}$$

or: $$\log_e N = 2.3026 \times \log_{10} N$$

Hence we may find the natural logarithm of a number by first finding its common logarithm and then multiplying by 2.302 6 (or dividing by 0.434 3).

EXAMPLE 3

By using common logarithms find the value of $\log_e 25$.

We have: $$\log_e N = 2.302\,6 \times \log_{10} N$$

$\therefore$ $$\log_e 25 = 2.302\,6 \times \log_{10} 25$$

$$= 2.302\,6 \times 1.397\,9$$

$$= 3.218\,9$$

EXAMPLE 4

By using common logarithms find $\log_e 0.396\,4$.

We have: $$\log_e 0.396\,4 = 2.302\,6 \times \log_{10} 0.396\,4$$

$$= 2.302\,6 \times \bar{1}.598\,1$$

$$= 2.302\,6(-1 + 0.598\,1)$$

$$= 2.302\,6(-0.401\,9)$$

$$= -0.925\,4$$

EXAMPLE 5

Find the number whose natural logarithm is 0.631.

If we let the required number be N then:

$$\log_e N = 0.631$$

but we know that: $\log_e N = 2.302\,6 \times \log_{10} N$

$\therefore$ $0.631 = 2.302\,6 \times \log_{10} N$

or: $\log_{10} N = \dfrac{0.631}{2.302\,6}$

$= 0.274\,0$

$\therefore$ $N = \text{antilog}_{10}\,0.274\,0$

$= 1.879$

Problems involving the use of natural logarithms occur frequently in electrical engineering. If a scientific electronic calculator is available the use of tables may be avoided. Remember the natural logarithm can be found by using the log$_e$ (or ln) key and the natural antilogarithm by using the e^x key. In the examples which follow two alternative solutions are given — one using tables and the other using a calculating machine.

EXAMPLE 6

Evaluate $50 \cdot e^{2.16}$.

(a) Calculations using common logarithms:

Let: $x = 50 \cdot e^{2.16}$

and taking common logarithms of both sides:

$\log_{10} x = \log_{10} 50 \cdot e^{2.16}$

$= \log_{10} 50 + \log_{10} e^{2.16}$

$= \log_{10} 50 + 2.16 \times \log_{10} e$

$= 1.699\,0 + 2.16 \times 0.434\,3$

$= 2.637\,1$

$\therefore$ $x = \text{antilog}_{10}\,2.637\,1$

$\therefore$ $x = 433.6$

(b) Using a calculating machine the sequence of operations would be:

$\boxed{C}\ \boxed{2}\ \boxed{\cdot}\ \boxed{1}\ \boxed{6}\ \boxed{e^x}\ \boxed{\times}\ \boxed{5}\ \boxed{0}\ \boxed{=}$

giving an answer 433.6.

EXAMPLE 7

Evaluate $200 \cdot e^{-1.34}$

(a) Calculations using common logarithms:

Let: $\qquad x = 200 \cdot e^{-1.34}$

and taking common logarithms of both sides:

$$\log_{10} x = \log_{10} 200 \cdot e^{-1.34}$$
$$= \log_{10} 200 + \log_{10} e^{-1.34}$$
$$= \log_{10} 200 + (-1.34) \times \log_{10} e$$
$$= 2.301\,0 - 1.34 \times 0.434\,3$$
$$= 1.719\,0$$
$$\therefore \qquad x = \text{antilog}_{10}\, 1.719\,0$$
$$\therefore \qquad x = 52.36$$

(b) Using a calculating machine:

Here we need to enter first a negative number, -1.34.

Procedures differ on various types of machine — we will assume that the number is entered first as an ordinary positive value and the sign changed by using the 'change sign' key,

The sequence of operations would then be:

$$\boxed{C}\,\boxed{1}\,\boxed{\cdot}\,\boxed{3}\,\boxed{4}\,\boxed{\substack{\text{chg}\\\text{sign}}}\,\boxed{e^x}\,\boxed{\times}\,\boxed{2}\,\boxed{0}\,\boxed{0}\,\boxed{=}$$

giving an answer 52.37.

EXAMPLE 8

In a capacitive circuit the instantaneous voltage across the capacitor is given by $v = V(1 - e^{-t/CR})$ where V is the initial supply voltage, R ohms the resistance, C farads the capacitance, and t seconds the time from the instant of connecting the supply voltage.

If $V = 200$, $R = 10\,000$, and $C = 20 \times 10^{-6}$ find the time when the voltage v is 100 volts.

Substituting the given values in the equation we have:

$$100 = 200(1 - e^{-t/20 \times 10^{-6} \times 10\ 000})$$

$$\therefore \qquad \frac{100}{200} = 1 - e^{-t/0.2}$$

$$\therefore \qquad 0.5 = 1 - e^{-5t}$$

$$\therefore \qquad e^{-5t} = 1 - 0.5$$

$$\therefore \qquad e^{-5t} = 0.5$$

(a) Calculations using common logarithms:

Taking logarithms of both sides:

$$\log_{10} e^{-5t} = \log_{10} 0.5$$

$$\therefore \qquad -5t(\log_{10} e) = \log_{10} 0.5$$

$$\therefore \qquad -5t(0.434\ 3) = \bar{1}.699\ 0$$

$$\therefore \qquad -5t = \frac{-1 + 0.699\ 0}{0.434\ 3}$$

$$\therefore \qquad -5t = \frac{-0.301\ 0}{0.434\ 0}$$

$$\therefore \qquad -5t = -0.693\ 1$$

$$\therefore \qquad t = \frac{0.693\ 1}{5}$$

$$\therefore \qquad t = 0.138\ 6 \text{ seconds}$$

(b) Calculations using a calculating machine:

Rewriting: $\qquad\qquad e^{-5t} = 0.5$

in logarithmic form we have

$$\log_e 0.5 = -5t$$

$$\therefore \qquad t = -\frac{\log_e 0.5}{5}$$

and the sequence of operations would then be:

giving an answer 0.138 6 seconds.

EXAMPLE 9

$$R = \frac{(0.42)S}{l} \times \log_e \frac{d_2}{d_1}$$

refers to the insulation resistance of a wire. Find the value of R when $S = 2\,000$, $l = 120$, $d_1 = 0.2$ and $d_2 = 0.3$.

Substituting the given values,

$$R = \frac{0.42 \times 2000}{120} \times \log_e \frac{0.3}{0.2}$$

$$= \frac{0.42 \times 2000}{120} \times \log_e 1.5$$

(a) Calculations using common logarithms i.e. $\log_{10}$:

$$R = \frac{0.42 \times 2000}{120} \, 2.302\,6 \times \log_{10} 1.5$$

$$= \frac{0.42 \times 2000 \times 2.302\,6 \times 0.176\,1}{120}$$

$$= 2.839$$

Number	$\log_{10}$
0.42	$\overline{1}.623\,2$
2 000	$3.301\,0$
0.176 1	$\overline{1}.245\,8$
2.302 6	$0.362\,3$
Numerator	$2.532\,3$
120	$2.079\,2$
2.839	$0.453\,1$

(b) Calculations using a machine:

The sequence of operations would be:

$$\boxed{C}\,\boxed{1}\,\boxed{\cdot}\,\boxed{5}\,\boxed{\ln}\,\boxed{\times}\,\boxed{0}\,\boxed{\cdot}\,\boxed{4}\,\boxed{2}\,\boxed{\times}\,\boxed{2}\,\boxed{0}\,\boxed{0}\,\boxed{0}\,\boxed{\div}\,\boxed{1}\,\boxed{2}\,\boxed{0}\,\boxed{=}$$

giving an answer 2.838.

Exercise 9

1) Use tables of natural logarithms to find the values of:

(a) $\log_e 311.3$ (b) $\log_e 7.081$ (c) $\log_e 20.15$
(d) $\log_e 8.128$ (e) $\log_e 0.491\,3$ (f) $\log_e 0.008\,178$

2) By using common logarithms find the values of:

(a) $\log_e 36$ (b) $\log_e 482$ (c) $\log_e 0.955\,4$
(d) $\log_e 0.038\,1$ (e) $\log_e 2.07$ (f) $\log_e 0.10$

3) Find the numbers whose natural logarithms are:

(a) 2.76   (b) 0.677   (c) 0.09

(d) −3.46   (e) −0.543   (f) −0.078

4) Find the values of:

(a) $70e^{2.5}$ (b) $150e^{-1.34}$ (c) $3.4e^{-0.445}$

5) The formula

$$L = 0.000\,644 \left\{ \log_e \frac{d}{r} + \frac{1}{4} \right\}$$

is used for calculating the self-inductance of parallel conductors. Find L when $d = 50$ and $r = 0.25$.

6) The inductance (L microhenrys) of a straight aerial is given by the formula:

$$L = \frac{1}{500} \left(\log_e \frac{4l}{d} - 1 \right)$$

where l is the length of the aerial in cm and d its diameter in cm. Calculate the inductance of an aerial 500 cm long and 0.20 cm in diameter.

7) Find the value of $\log_e \left(\frac{c_1}{c_2} \right)^2$ when $c_1 = 4.7$ and $c_2 = 3.5$.

8) If $T = R \log_e \left(\frac{a}{a-b} \right)$

find T when $R = 28$, $a = 5$ and $b = 3$.

9) When a chain of length $2l$ is suspended from two points $2d$ apart on the same horizontal level,

$$d = c \log_e \left(\frac{l + \sqrt{l^2 + c^2}}{c} \right)$$

If $c = 80$ and $l = 200$ find d.

10) The instantaneous value of the current when an inductive circuit is discharging is given by the formula $i = Ie^{-Rt/L}$. Find the value of this current, i, when $R = 30$, $L = 0.5$ and $t = 0.005$.

11) In a circuit in which a resistor is connected in series with a capacitor the instantaneous voltage across the capacitor is given by the formula $v = V(1 - e^{-t/CR})$. Find this voltage, v, when $V = 200$, $C = 40 \times 10^{-6}$, $R = 100\,000$ and $t = 1$.

12) In the formula $v = Ve^{-Rt/L}$ the values of v, V, R and L are 50, 150, 60 and 0.3 respectively. Find the corresponding value of t.

13) The instantaneous charge in a capacitive circuit is given by $q = Q$ $(1 - e^{-t/CR})$. Find the value of t when $q = 0.01$, $Q = 0.015$, $C = 0.0001$, and $R = 7000$.

SUMMARY

Natural logarithms are also called Hyperbolic or Naperian logarithms and are denoted by $\log_e$ or ln.

$$\log_e N = 2.302\,6 \times (\log_{10} N)$$

6. GRAPHS

After reaching the end of this chapter you should be able to :-

1. State that the equation of a straight line is $y = mx + c$.

2. Convert expressions which, with direct plotting produce curves on a graphical field, but which with appropriate axes, produce straight lines.

AXES OF REFERENCE

To plot a graph we first draw two lines at right angles to each other (Fig. 6.1). These lines are called the axes of reference. The horizontal axis is often called the x-axis and the vertical axis the y-axis.

CO-ORDINATES

Co-ordinates are used to mark the points on a graph. In Fig. 6.1 the point P has been plotted so that $x = 1$ and $y = 5$. The values 1 and 5 are said to be the rectangular co-ordinates of P. For brevity we say that P is the point $(1, 5)$.

In plotting graphs we may have to include co-ordinates which are positive and negative. To represent these on a graph we make use of the number scales used in directed numbers. As well as the point $(1, 5)$ the points $(3, -15)$, $(-2, 10)$ and $(-3, -10)$ are plotted in Fig. 6.1.

The Origin

If the zero of both axes occurs at the intersection of the axes as in Fig. 6.1, then this point $(0, 0)$ is called the *origin*.

AXES AND SCALES

The location of the axes and the scales along each axis should be chosen so that all the points may be plotted with the greatest possible accuracy. The scales should be as large as possible but they must be chosen so that they are easy to read. The most useful scales are 1, 2 and 5 units to 1 large square on the graph paper. Some multiples of these such as 10, 20, 100 units etc. per large square are also suitable. Note that the scales chosen need not be the same on both axes.

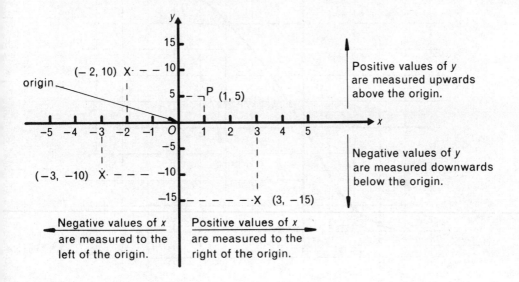

Fig. 6.1

EXAMPLE 1

The table below gives corresponding values of x and y. Plot this information and from the graph find:

a) the value of y when $x = -3$

b) the value of x when $y = 2$

x	-4	-2	0	2	4	6
y	-2.0	-1.6	0	1.4	2.5	3.0

The graph is shown plotted in Fig. 6.2 and it is a smooth curve. This means that there is a definite law (or equation) connecting x and y. We can therefore use the graph to find corresponding values of x and y between those given in the original table of values. By using the constructions shown in Fig. 6.2:

a) the value of y is -1.9 when $x = -3$;

b) the value of x is 3 when $y = 2$.

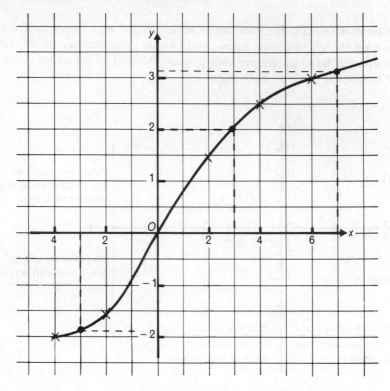

Fig. 6.2

Using a graph in this way to find values of x and y not given in the original table of values is called *interpolation*. If we extend the curve so that it follows the general trend we can estimate corresponding values of x and y which lie *just beyond* the range of the given values. Thus in Fig. 6.2 by extending the curve we can find the probable value of y when $x = 7$. This is found to be 3.2. Finding a probable value in this way is called *extrapolation*. An extrapolated value can usually be relied upon but in some cases it may contain a substantial amount of error. Extrapolated values must therefore be used with care.

It must be clearly understood that interpolation and extrapolation can only be used if the graph is a straight line or a smooth curve.

EXAMPLE 2

Corresponding values of x and y are shown in the table below

x	0	10	20	30	40	50
y	20.0	22.0	23.5	24.4	25.0	25.4

Illustrate this relationship on a graph.

Looking at the range of values for y, we see that they range from 20.0 to 25.4. We can therefore make 20.0 the starting point on the vertical axis as shown in Fig. 6.3. By doing this a larger scale may be used on the y-axis thus resulting in a more accurate graph. The graph is again a smooth curve and hence there is a definite equation connecting x and y.

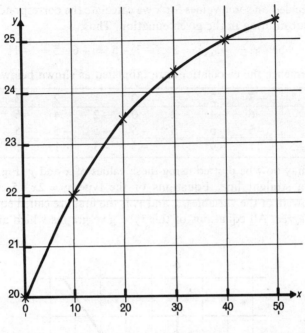

Fig. 6.3

GRAPHS OF SIMPLE EQUATIONS

Consider the equation:

$$y = 2x+5$$

We can give x any value we please and so calculate a corresponding value for y. Thus,

when $x = 0$ $y = 2\times0+5 = 5$
when $x = 1$ $y = 2\times1+5 = 7$
when $x = 2$ $y = 2\times2+5 = 9$ and so on

The value of y therefore depends on the value allocated to x. We therefore call y the *dependent variable*. Since we can give x any value we please, we call x the *independent variable*. It is usual to mark the values of the independent variable along the horizontal x-axis and the values of the dependent variable are then marked off along the vertical y-axis.

EXAMPLE 3

Draw the graph of $y = 2x - 5$ for values of x between -3 and 4.

Having decided on some values for x we calculate the corresponding values for y by substituting in the given equation. Thus,

$$\text{when } x = -3, y = 2 \times (-3) - 5 = -6 - 5 = -11$$

For convenience the calculations are tabulated as shown below.

x	-3	-2	-1	0	1	2	3	4
$2x$	-6	-4	-2	0	2	4	6	8
-5	-5	-5	-5	-5	-5	-5	-5	-5
$y = 2x - 5$	-11	-9	-7	-5	-3	-1	1	3

A graph may now be plotted using these values of x and y (Fig. 6.4). The graph is a straight line. Equations of the type $y = 2x - 5$, where the highest powers of the variables, x and y, is the first are called equations of the *first degree*. All equations of this type give graphs which are straight

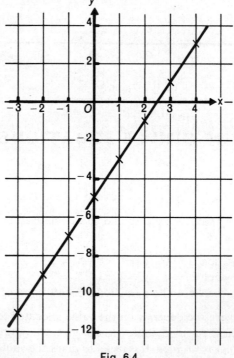

Fig. 6.4

lines and hence they are often called *linear equations*. In order to draw graphs of linear equations we need only take two points. It is safer, however, to take three points, the third point acting as a check on the other two.

EXAMPLE 4

By means of a graph show the relationship between x and y in the equation $y = 5x+3$. Plot the graph between $x = -3$ and $x = 3$.

Since this is a linear equation we need only take three points.

x	-3	0	$+3$
$y = 5x+3$	-12	3	$+18$

The graph is shown in Fig. 6.5.

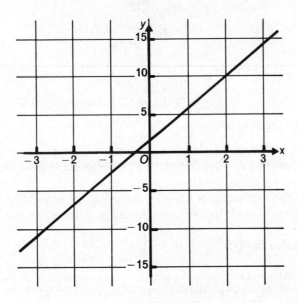

Fig. 6.5

THE LAW OF A STRAIGHT LINE

In Fig. 6.6, the point B is any point on the line shown and has co-ordinates x and y. Point A is where the line cuts the y-axis and has co-ordinates $x = 0$ and $y = c$.

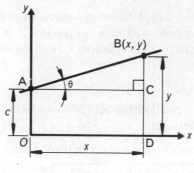

Fig. 6.6

In $\triangle ABC$

$$\frac{BC}{AC} = \tan \theta$$

$\therefore$

$$BC = (\tan \theta).AC$$

but also

$$y = BC + CD$$

$$= (\tan \theta).AC + CD$$

$\therefore$

$$y = mx + c$$

where

$$m = \tan \theta$$

and

$$c = \text{distance } CD = \text{distance } OA$$

m is called the *gradient of the line*.

c is called the *intercept on the y-axis*. Care must be taken as this only applies if the origin (i.e. the point (0, 0)) is at the intersection of the axes.

In mathematics the gradient of a line is defined as the tangent of the angle that the line makes with the horizontal, and is denoted by the letter m.

Hence in Fig. 6.6 the gradient $= m = \tan \theta = \dfrac{BC}{AC}$

(Care should be taken not to confuse this with the gradient given on maps, railways, etc. which is the sine of the angle (not the tangent) — e.g. a railway slope of 1 in 100 is one unit vertically for every 100 units measured along the slope.)

Fig. 6.7 shows the difference between positive and negative gradients.

Fig. 6.7

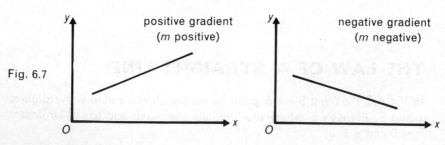

Summarising:

The standard equation, or law, of a straight line is $y = mx + c$

where m is the gradient

and c is the intercept on the y-axis.

OBTAINING THE STRAIGHT LINE LAW OF A GRAPH

Two methods are used:

(i) Origin at the intersection of the axes

When it is convenient to arrange the origin, i.e. the point $(0, 0)$, at the intersection of the axes the values of gradient m and intercept c may be found directly from the graph as shown in Example 5.

EXAMPLE 5

Find the law of the straight line shown in Fig. 6.8.

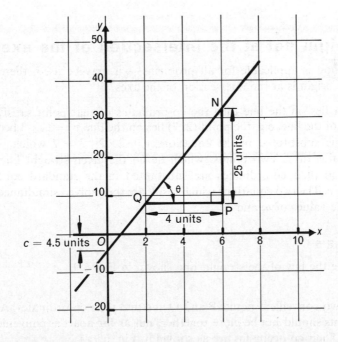

Fig. 6.8

To find gradient *m*. Take any two points Q and N on the line and construct the right angled triangle QPN. This triangle should be of reasonable size, since a small triangle will probably give an inaccurate result. Note that if we can measure to an accuracy of 1 mm using an ordinary rule, then this error in a length of 20 mm is much more significant than the same error in a length of 50 mm.

The lengths of NP and QP are then found using the scales of the x and y axes. Direct lengths of these lines as would be obtained using an ordinary rule, e.g. both in centimetres, must *not* be used — the scales of the axes must be taken into account.

$$\therefore \qquad \text{gradient } m = \tan \theta = \frac{NP}{QP} = \frac{25}{4} = 6.25$$

To find intercept *c*. This is measured again using the scale of the y-axis.

$\therefore$ intercept $\qquad\qquad c = -4.5$

The law of the straight line.

The standard equation is $\quad y = mx + c$

$\therefore$ the required equation is $\ y = 6.25x + (-4.5)$

i.e. $\qquad\qquad\qquad\qquad y = 6.25x - 4.5$

(ii) Origin not at the intersection of the axes

This method is applicable for all problems — it may be used, therefore, when the origin is at the intersection of the axes.

If a point lies on the line then the co-ordinates of that point satisfy the equation of the line, e.g. the point (2, 7) lies on the line $y = 2x + 3$ because if $x = 2$ is substituted in the equation, $y = 2 \times 2 + 3 = 7$ which is the correct value of y. Two points, which lie on the given straight line, are chosen and their co-ordinates are substituted in the standard equation $y = mx + c$. The two equations which result are then solved simultaneously to find the values of m and c.

EXAMPLE 6

Determine the law of the straight line shown in Fig. 6.9.

Choose two convenient points P and Q and find their co-ordinates. Again these points should not be close together, but as far apart as conveniently possible. Their co-ordinates are as shown in Fig. 6.9.

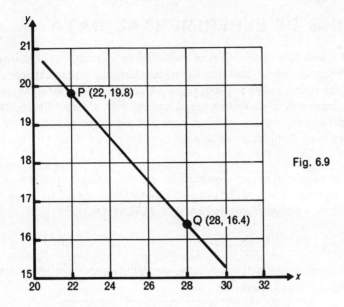

Fig. 6.9

Let the equation of the line be $y = mx + c$

Now P (22, 19.8) lies on the line $\therefore$ $19.8 = m(22) + c$

and Q (28, 16.4) lies on the line $\therefore$ $16.4 = m(28) + c$

To solve these two equations simultaneously we must first eliminate one of the unknowns. In this case c will disappear if the second equation is subtracted from the first, giving

$$19.8 - 16.4 = m(22 - 28)$$

i.e.

$$3.4 = m(-6)$$

$\therefore$

$$m = \frac{3.4}{-6}$$

$\therefore$

$$m = -0.567$$

To find c the value of $m = -0.567$ may be substituted into either of the original equations. Choosing the first equation we get

$$19.8 = -0.567(22) + c$$

i.e.

$$19.8 = -12.47 + c$$

$\therefore$

$$c = 19.8 + 12.47$$

$\therefore$

$$c = 32.27$$

Hence the required law of the straight line is

$$y = -0.567x + 32.27$$

GRAPHS OF EXPERIMENTAL DATA

Readings which are obtained as a result of an experiment will usually contain errors owing to inaccurate measurement and other experimental errors. If the points, when plotted, show a trend towards a straight line or a smooth curve this is usually accepted and the best straight line or curve drawn. In this case the line will not pass through some of the points and an attempt must be made to ensure an even spread of these points above and below the line or the curve.

One of the most important applications of the straight line law is the determination of a law connecting two quantities when values have been obtained from an experiment as Example 7 illustrates.

EXAMPLE 7

A test on a metal-filament lamp gave the following values of resistance (R ohms) at various voltages (V volts).

V	62	75	89	100	120
R	100	118	136	149	176

These results are expected to agree with a law of the type $R = aV+b$ where a and b are constants. Test that this is so by drawing the graph of resistance (plotted vertically) against voltage (plotted horizontally) and find the law. Hence find the value of V when R is 142 ohms.

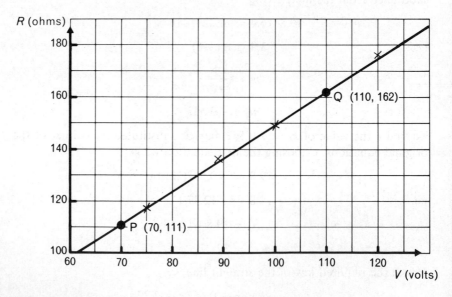

Fig. 6.10

The standard equation of a straight line is $y = mx+c$. It often happens that the variables are *not* x and y. In this example V is used instead of x and is plotted on the horizontal axis, and R is used instead of y and is plotted on the vertical axis.

Similarly the gradient is a instead of m, and the intercept on the y-axis is b instead of c.

The points are shown plotted in Fig. 6.10. If V and R are connected by a law of the type $R = aV+b$ then the graph must be a straight line. The points deviate slightly from a straight line but because the data are experimental, errors in measurement and observation must be expected and hence slight deviations from a straight line are bound to occur. Therefore we can say $R = mV+c$.

To find the values of m and c choose two points such as P and Q which lie on the graph and which are not too close together. Read off their co-ordinates which are:

at P, $V = 70$ and $R = 111$

at Q, $V = 110$ and $R = 162$

Substituting these values in the equation $R = mV+c$:

$$162 = 110m+c \tag{1}$$

$$111 = 70m+c \tag{2}$$

Subtracting equation (2) from equation (1),

$$51 = 40m$$

$$m = \frac{51}{40} = 1.28$$

Substituting $m = 1.28$ in equation (1),

$$162 = 110 \times 1.28+c$$

$$162 = 140.8+c$$

$$\therefore \qquad c = 21.2$$

Hence: $$R = 1.28V+21.2$$

To find V when $R = 142$ ohms, the value $R = 142$ is substituted into the equation $R = 1.28V+21.2$ giving:

$$142 = 1.28V+21.2$$

$$\therefore \qquad V = \frac{142 - 21.2}{1.28}$$

$$\therefore \qquad V = 94.4 \text{ volts}$$

(since all values of V are volts when values of R are ohms).

This value of V may be verified by checking the value of V corresponding to $R = 142$ on the straight line in Fig. 6.10.

Exercise 10

1) Draw the straight line which passes through the points $(4, 7)$ and $(-2, 1)$. Hence find the gradient of the line and its intercept on the y axis.

2) The following equations all represent straight lines. For each equation state the gradient and the intercept on the y axis.

(a) $y = x + 3$ (b) $y = -3x + 4$
(c) $y = -31x - 1.7$ (d) $y = 4.3x - 2.5$

3) A straight line passes through the points $(-2, -3)$ and $(3, 7)$. Without drawing the line find the values of m and c in the equation $y = mx + c$.

4) The following table gives values of x and y which are connected by a law of the type $y = mx + c$. Plot the graph and from it find the values of m and c.

x	2	4	6	8	10	12
y	10	16	22	28	34	40

5) The following observed values of P and Q are supposed to be related by a law of the type $P = aQ + b$, but there are experimental errors. Find by plotting P and Q the most probable values of a and b.

Q	2.5	3.5	4.4	5.8	7.5	9.6	12.0	15.1
P	13.6	17.6	22.2	28.0	35.5	47.4	56.1	74.6

6) During a test with a thermo-couple pyrometer the e.m.f. (E millivolts) was measured against the temperature at the hot junction (t °C) and the following results obtained:

t	200	300	400	500	600	700	800	900	1000
E	6	9.1	12.0	14.8	18.2	21.0	24.1	26.8	30.2

The law connecting t and E is supposed to be $E = at + b$. Test if this is so and find suitable values for a and b.

7) The resistance (R ohms) of a field winding is measured at various temperatures (t °C) and the results recorded in the table opposite.

t (°C)	21	26	33	38	47	54	59	66	75
R (ohms)	109	111	114	116	120	123	125	128	132

If the law connecting R and t is of the form $R = a + bt$ find suitable values of a and b.

8) A test on a metal filament lamp gave the following values of resistance (R ohms) at various voltages (V volts).

V	62	75	89	100	120
R	100	117	135	149	175

These results are expected to agree with a law of the type $R = mV + c$, where m and c are constants. Test this by drawing the graph and from the graph find suitable values for m and c.

9) The fall of potential along a uniform wire was measured with the results shown below:

Length of wire (l cm)	20	40	60	80	100
Fall of potential (V volts)	0.4	0.78	1.20	1.56	1.95

These results are expected to agree with a law of the type $V = kl$ where k is a constant. By drawing a graph show that the law is of the type suggested and from it find the value of k.

10) The critical grid voltages V_g of an electronic component, for values of anode voltage V_a are given below.

V_a	10	30	60	100	170	240	330	440	500
V_g	0	-2	-3.5	-5	-7.5	-10	-13	-17	-19

Plot a graph of V_a (y-axis) against V_g (x-axis). If the law connecting V_a and V_g for the straight line portion of the graph is $V_a = mV_g + c$, determine the values of the constants m and c.

NON-LINEAR LAWS WHICH CAN BE REDUCED TO THE LINEAR FORM

Many non-linear equations can be reduced to the linear form by making a suitable substitution.

Common forms of non-linear equations are:

$$y = \frac{a}{x} + b$$

$$y = \frac{a}{x^2} + b$$

$$y = ax^2 + b$$
$$y = a\sqrt{x} + b$$
$$y = ax^2 + bx$$

Where in each case a and b are constants.

Consider $y = \dfrac{a}{x} + b$

Let $z = \dfrac{1}{x}$ so that the equation becomes $y = az + b$. If we now plot values of y against the corresponding values of z we will get a straight line since $y = az + b$ is of the standard linear form. In effect y has been plotted against $\dfrac{1}{x}$.

The following example illustrates this method:

EXAMPLE 8

The voltage V across the arc of a carbon filament lamp for values of the current I in the arc were measured in an experiment and the results are shown in the following table:

I	1.0	1.5	2.0	2.5	3.0	3.5	4.0
V	82.0	68.7	62.0	58.0	55.3	53.4	52.0

The relation between V and I is thought to be of the form $V = \dfrac{a}{I} + b$.

Check this and find the values of a and b.

By putting $z = \dfrac{1}{I}$ the equation will be reduced to $V = az + b$ which is a straight line equation. To try this we shall draw up a table of V and z, and plot the values obtained.

I	1.0	1.5	2.0	2.5	3.0	3.5	4.0
$z = \dfrac{1}{I}$	1.0	0.667	0.500	0.400	0.333	0.286	0.250
V	82.0	68.7	62.0	58.0	55.3	53.4	52.0

The graph obtained is shown in Fig. 6.11, and since it is a straight line it follows the law is of the form $V = az + b$, i.e. of the form $V = \dfrac{a}{I} + b$.

To find the values of a and b the two point method must be used since the *origin* is *not* at the intersection of the axes.

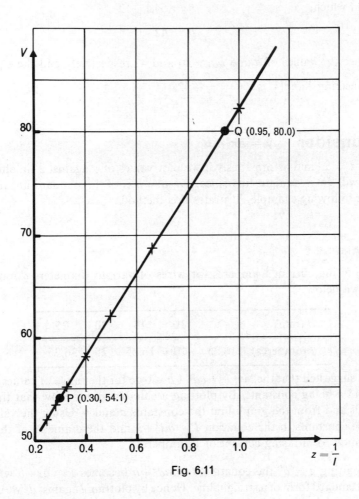

Fig. 6.11

The point (0.30, 54.1) lies on the line $\quad\therefore\ 54.1 = a(0.30) + b$

and the point (0.95, 80.0) lies on the line $\quad\therefore\ 80.0 = a(0.95) + b$

hence subtracting the first equation from the second we get,

$$80.0 - 54.1 = a(0.95 - 0.30)$$

i.e. $$25.9 = a(0.65)$$

$$\therefore \quad a = \frac{25.9}{0.65} = 40$$

Substituting this value in the first equation we obtain:

$$54.1 = 40(0.30) + b$$

from which

$$b = 54.1 - 12$$

∴

$$b = 42$$

Hence the values of a and b are 40 and 42 respectively and the equation connecting I and V is $V = \dfrac{40}{I} + 42$.

Consider $y = ax^2 + b$.

Let $z = x^2$ and as previously if we plot values of y against z (in effect x^2) we will get a straight line since $y = az + b$ is of the standard linear form. The following example illustrates this method:

EXAMPLE 9

The fusing current I amperes for wires of various diameters d mm is as shown below:

d (mm)	5	10	15	20	25
I (amperes)	6.25	10	16.25	25	36.25

It is suggested that the law $I = ad^2 + b$ is true for the range of values given, a and b being constants. By plotting a suitable graph show that this law holds and from the graph find the constants a and b. Using the values of these constants in the equation $I = ad^2 + b$ find the diameter of the wire required for a fusing current of 12 amperes.

By putting $z = d^2$ the equation $I = ad^2 + b$ becomes $I = az + b$ which is the standard form of a straight line. Hence by plotting I against d^2 we should get a straight line if the law is true. To try this we draw up a table showing corresponding values of I and d^2.

d	5	10	15	20	25
$z = d^2$	25	100	225	400	625
I	6.25	10	16.25	25	36.25

From the graph (Fig. 6.12) we see that the points do lie on a straight line and hence the values obey a law of the form $I = ad^2 + b$.

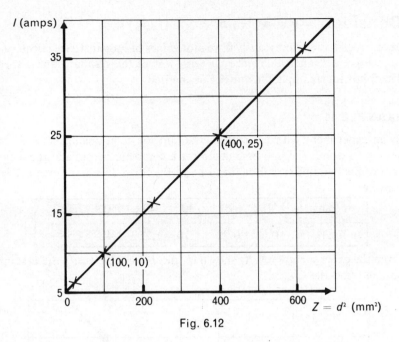

Fig. 6.12

To find the values of a and b choose two points which lie on the line and find their co-ordinates.

The point (400, 25) lies on the line, $\therefore$ $\quad 25 = 400a + b$ $\qquad$ (1)

The point (100, 10) lies on the line, $\therefore$ $\quad 10 = 100a + b$ $\qquad$ (2)

Subtracting equation (2) from equation (1),

$$15 = 300a$$
$$a = 0.05$$

Substituting $a = 0.05$ in equation (2),

$$10 = 100 \times 0.05 + b$$
$$b = 5$$

Therefore the law is:

$$I = 0.05d^2 + 5$$

When $I = 12$,

$$12 = 0.05d^2 + 5$$
$$0.05d^2 = 7$$
$$d^2 = \frac{7}{0.05} = 140$$
$$d = \sqrt{140} = 11.83 \text{ mm}$$

Consider $\quad y = a\sqrt{x} + b.$

Let $z = \sqrt{x}$ and as previously if we plot values of y against z (in effect $\sqrt{x}$) we will get a straight line since $y = az + b$ is of the standard linear form. The following example illustrates this method:

EXAMPLE 10

In an experiment with a simple pendulum the relationship between the period T seconds (i.e. the time taken for a complete swing) and the length l cm of the pendulum is thought to be of the form $T = k\sqrt{l}$ where k is a constant.

l (cm)	10	20	30	40	50	60
T (s)	0.62	0.90	1.10	1.25	1.42	1.55

From the experimental results shown in the above table verify the assumption and find the law.

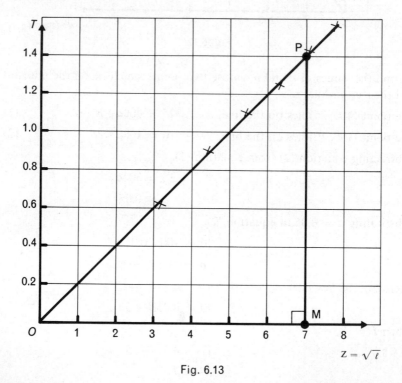

Fig. 6.13

By putting $z = \sqrt{l}$ the assumed equation becomes $T = kz$. Hence by plotting T against z we should get a straight line if the results follow the law with the constant $b = 0$.

The values of T and z are shown in the following table:

T	0.62	0.90	1.10	1.25	1.42	1.55
$z = \sqrt{l}$	3.16	4.47	5.47	6.32	7.07	7.75

The graph of these results is shown in Fig. 6.13.

Within experimental error the points lie on a straight line verifying a linear relationship. Furthermore since the *origin* is shown at the intersection of the axes then the intercept b is given by where the line cuts the vertical axis. This is zero since the line passes through the origin (i.e. the point 0, 0).

To find a gradient a, construct a reasonably sized triangle OPM.

Then the gradient $a = \dfrac{PM}{OM} = \dfrac{1.39}{7} = 0.20$

Hence the law is verified and is $T = 0.20\sqrt{l}$

Consider $y = ax^2 + bx$.

Dividing both sides by the equation, by x we get $\dfrac{y}{x} = ax + b$. Now let

$t = \dfrac{y}{x}$ so that the equation now becomes $t = ax + b$. If we now plot values of t against corresponding values of x we will get a straight line since $t = ax + b$ is of the standard linear form. In effect $\dfrac{y}{x}$ has been plotted against x.

The following example illustrates this method:

EXAMPLE 11

Verify that the following table of values satisfies the equation $y = ax^2 + bx$ and find the values of a and b.

x	2	3	4	5	6
y	22	45	76	115	162

We have $y = ax^2 + bx$.

Dividing both sides by x,

$$\frac{y}{x} = ax + b$$

If we put $t = \dfrac{y}{x}$ then,

$$t = ax + b$$

which is of the standard form. Hence if we plot $\dfrac{y}{x}$ against x we should get a straight line if the assumed law $y = ax^2 + bx$ is correct. Drawing up a table of x against $\dfrac{y}{x}$ we get:

x	2	3	4	5	6
y	22	45	76	115	162
$\dfrac{y}{x}$	11	15	19	23	27

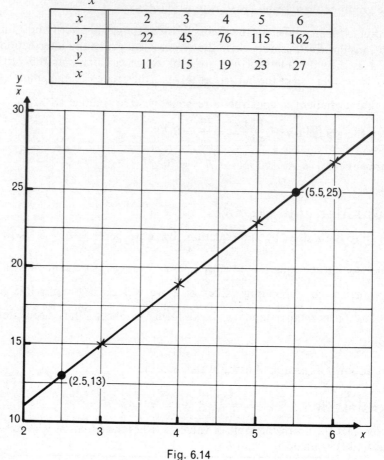

Fig. 6.14

From the graph (Fig. 6.14) we see that the points do, in fact, lie on a straight line and the given values therefore obey the law:

$$\frac{y}{x} = ax + b$$

or:

$$y = ax^2 + bx$$

To find the values of the constants a and b we choose two suitable points and find their co-ordinates.

Point (5.5, 25) lies on the line,

$\therefore$
$$25 = 5.5a + b \qquad (1)$$

Point (2.5, 13) lies on the line,

$$\therefore \qquad 13 = 2.5a + b \qquad\qquad (2)$$

Subtracting equation (2) from equation (1),

$$12 = 3a$$

$$a = 4$$

Substituting $a = 4$, in equation (2),

$$13 = 2.5 \times 4 + b$$

$$b = 3$$

Hence the law becomes:

$$y = 4x^2 + 3x$$

Exercise 11

1) Corresponding values obtained experimentally for two quantities are:

x	1	2	3	4	5
y	4.9	11.1	21.0	34.1	52.9

A law of the form $y = ax^2 + b$ is suspected. By plotting y vertically against x^2 horizontally prove that the law is as stated. Hence find values for a and b.

2) The values tabulated below are thought to obey the law $y = m\sqrt{x} + c$. By plotting y (vertically) against $\sqrt{x}$ (horizontally) show that this is so and hence find values for m and c.

x	1	4	9	16	25
y	2	3.5	5	6.5	8

3) The following values of x and y obey the law $y = ax^3 + b$. By plotting a graph of y (vertically) against x^3 (horizontally) find the values of a and b.

x	1	2	3	4	5
y	0.3	1.0	2.9	6.6	12.7

4) The values tabulated below follow the law $y = \dfrac{a}{x} + b$. By plotting y (vertically) against $\dfrac{1}{x}$ (horizontally) prove the law and find suitable values for a and b.

x	0.1	0.2	0.4	0.5	1.0
y	31	16	8.5	7	4

5) The following values of x and y follow a law of the type $y = ax^2 + bx$. By plotting $\dfrac{y}{x}$ (vertically) against x (horizontally) find values for a and b.

x	4	6	8	10	12
y	11.2	19.2	28.8	40	52.8

6) The following readings were taken during a test:

R (ohms)	85	73.3	64	58.8	55.8
I (amperes)	2	3	5	8	12

R and I are thought to be connected by an equation of the form $R = \dfrac{a}{I} + b$.

Verify that this is so by plotting R (y-axis) against $\dfrac{1}{I}$ (x-axis) and hence find values for a and b.

7) In an experiment, the resistance R of copper wire of various diameters d mm was measured and the following readings obtained.

d mm	0.1	0.2	0.3	0.4	0.5
R ohms	20	5	2.2	1.3	0.8

Show that $R = \dfrac{k}{d^2}$ and find a suitable value for k.

8) The fusing current for different diameters of a certain wire is as shown below.

Diameter (x mm)	5	10	15	20	25
Fusing current (I amperes)	6.25	10	16.25	25	36.25

It is thought that $I = ax^2 + b$. By plotting a suitable graph show that this is so and hence find suitable values for a and b.

9) The total iron loss in a transformer is given by $P = k_1 f + k_2 f^2$ where k_1 and k_2 are constants. Using the following results plot a graph of $\dfrac{P}{f}$ against f and determine the values of the constants k_1 and k_2.

P watts	1.6	4.4	8.4	13.6	20.0	27.6
f hertz	10	20	30	40	50	60

10) Show that the following values of x and y follow a law of the type $y = ax^2 + bx^3$.

x	-1	1	3	5	7
y	0.1	0.3	4.5	17.5	44.1

What are the values of a and b.

SUMMARY

a) Co-ordinates are the values of x and y which fix a point on a graph with axes at right angles to each other — the point is denoted by (x, y).

b) The *origin* is the point $(0, 0)$.

c) When drawing a graph make full use of the space available on the sheet of graph paper.

d) Choose convenient scales, i.e. 1, 2, or 5 units (or these multiplied by powers of 10) per square of the graph paper.

e) A straight line graph may be called a linear graph. Two points only are necessary to fix a straight line, but a third point serves as a check.

f) The law of a straight line is $y = mx + c$ where m is the gradient and c is the intercept on the vertical axis (providing that the origin is at the intersection of the axes.)

g) The gradient m is the tangent of the angle the line makes with the horizontal.

positive gradient   negative gradient

h) If the origin is at the intersection of the axes then the gradient m may be found by drawing a suitable (i.e. reasonably large) triangle and dividing the vertical height by the horizontal length. The intercept c may be read directly from where the line cuts the vertical axis.

i) If the *origin* is *not* at the intersection of the axes then the two-point method must be used to determine m and c. (**N.B.** this method also holds if the origin is at the intersection of the axes).

j) The following equations represent graphs which may be reduced to a straight line form by making the corresponding substitutions:

$$y = \frac{a}{x} + b \qquad \text{substitute} \qquad z = \frac{1}{x}$$

$$y = \frac{a}{x^2} + b \qquad \text{substitute} \qquad z = \frac{1}{x^2}$$

$$y = ax^2 + b \qquad \text{substitute} \qquad z = x^2$$

$$y = a\sqrt{x}+b \quad \text{substitute} \quad z = \sqrt{x}$$

$y = ax^2+bx$

rearrange to $\dfrac{y}{x} = ax+b$ and substitute $t = \dfrac{y}{x}$

Self-Test 2

State which answer or answers are correct:

1) The origin is the point:

    a (1, 1)   **b** (0, 1)   **c** (1, 0)   **d** (0,0)

2) The recommended number of units per square when choosing a scale for a graph are:

    a 3   **b** 2   **c** 40   **d** 10

3) A line with a negative gradient is:

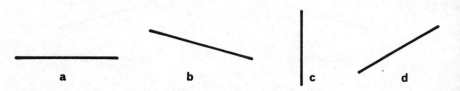

    a   **b**   **c**   **d**

4) Which of the following points lie on the line $y = 2x+3$?

    a (2, 1)   **b** $(-2, 0)$ **c** (2, 3)   **d** (2, 7)

5) Which of these equations represents a straight line?

    a $y = x^2+2$ **b** $y = 2x+3$ **c** $y = 7-x$ **d** $y^2 = x+1$

6) The gradient of the line $y = 4x+8$ is:

    a 8   **b** 1   **c** 2   **d** 4

7) The gradient of the line $y = 6-2x$ is:

    a 2   **b** 6   **c** -2 **d** -3

8) The values of gradient and intercept for the line $y = 3x$ are:

    a 1 and 3   **b** 3 and 1   **c** 1 and 0   **d** 3 and 0

9) The graph of $y = 2x+3$ will look like one of the diagrams in Fig. 6.15.

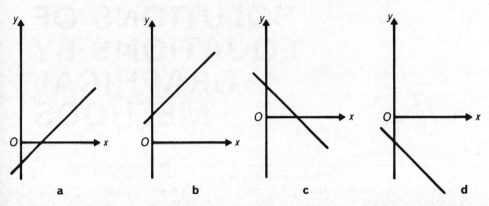

Fig. 6.15

10) The graph of $y = 5 - 3x$ will look like one of the following diagrams (Fig. 6.16).

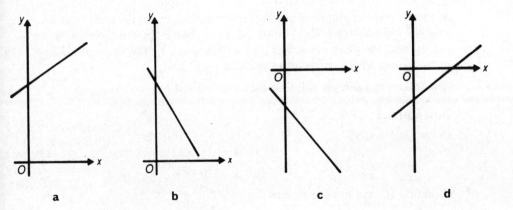

Fig. 6.16

SOLUTIONS OF EQUATIONS BY GRAPHICAL METHODS

After reaching the end of this chapter you should be able to:-

1. *Solve simultaneous equations in two unknowns by a graphical method.*
2. *Appreciate the effect of the constants a, b and c on the shape of a graph of a quadratic expression.*
3. *Solve a quadratic equation (a) by intersection the graph with the x-axis, (b) by intersection the graphs $y = ax^2$ and $y = ax+b$.*

SIMULTANEOUS LINEAR EQUATIONS

We first draw the graphs of the two given equations on the same axes and find the points where the graphs intersect. Since the solutions have to satisfy both the given equations they will be given by the value of x and y at the points where the graphs intersect.

The following example will illustrate the method.

EXAMPLE 1

Solve graphically:

$$y - 2x = 2 \tag{1}$$
$$3y + x = 20 \tag{2}$$

Equation (1) can be rewritten as:

$$y = 2x + 2$$

Equation (2) can be rewritten as:

$$y = \frac{20 - x}{3}$$

Drawing up the following table, we can plot the two equations on the *same axes.*

x	-3	0	3
$y = 2x+2$	-4	2	8
$y = \dfrac{20-x}{3}$	7.7	6.7	5.7

The solutions of the equations will be given by the co-ordinates of the point where the two lines cross (that is, point P in Fig. 7.1). The co-ordinates of P are $x = 2$ and $y = 6$. Hence the solutions are:

$$x = 2 \quad \text{and} \quad y = 6$$

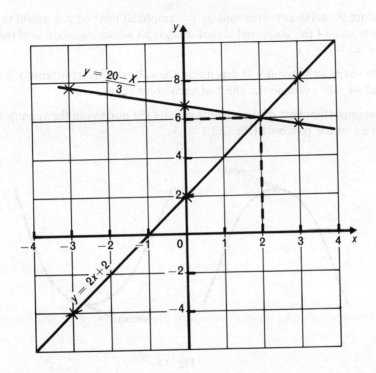

Fig. 7.1

Exercise 12

Solve graphically the following equations:

1) $3x + 2y = 7$
   $x + y = 3$

2) $4x - 3y = 1$
   $x + 3y = 19$

3) $x + 3y = 7$
   $2x - 2y = 6$

4) $2x - 3y = 5$
   $x - 2y = 2$

5) $7x - 4y = 37$
   $6x + 3y = 51$

6) $\dfrac{x}{2} + \dfrac{y}{3} = \dfrac{13}{6}$
   $\dfrac{2x}{7} - \dfrac{y}{4} = \dfrac{11}{6}$

QUADRATIC EQUATIONS

Quadratic equations may be solved by a graphical method which involves drawing a graph of the type $y = ax^2 + bx + c$. This is the equation of a curve called a parabola.

Before we solve any equations by the graphical method it is useful to have some idea of the shape and layout of a graph whose equation is of the type $y = ax^2 + bx + c$.

The shape and layout will depend on the values of the constants a, b and c and we will examine the effect of each constant in turn.

The important part of the curve is usually the portion in the vicinity of the vertex of the parabola (Fig. 7.2.)

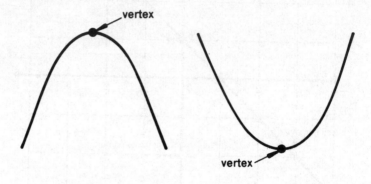

Fig. 7.2

In the examples which follow the graphs of various equations are shown. Tables of values have been omitted but the reader may find it useful to draw up tables of values and plot the curves himself.

Constant *a*

Fig. 7.3 shows the graphs of $y = ax^2$ when $a = 4$, $a = 2$ and $a = -1$.

We can see that if the value of *a* is positive the curve is shaped $\vee$, and the greater the value of *a* the 'steeper' the curve rises.

Negative values of *a* give a curve shaped $\wedge$.

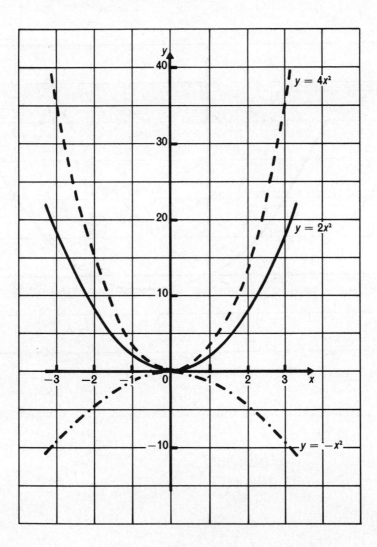

Fig. 7.3

Constant b

Fig. 7.4 shows the graphs of $y = x^2 + bx$ when $b = 2$ and $b = -3$. The effect of a positive value of b is to move the vertex to the left of the vertical y-axis, whilst a negative value of b moves the vertex to the right of the vertical axis.

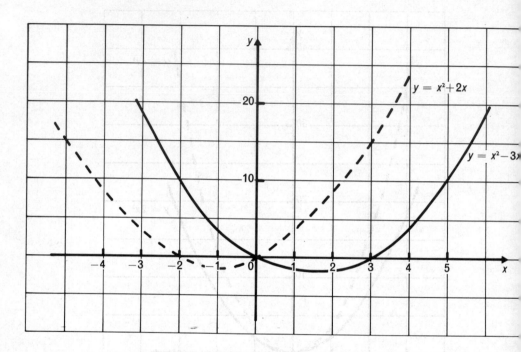

Fig. 7.4

Constant c

Fig. 7.5 shows the graphs of $y = x^2 - 2x + c$ when $c = 10$, $c = 5$, $c = 0$ and $c = -5$. As we can see the effect is to move the vertex up or down according to the magnitude of c.

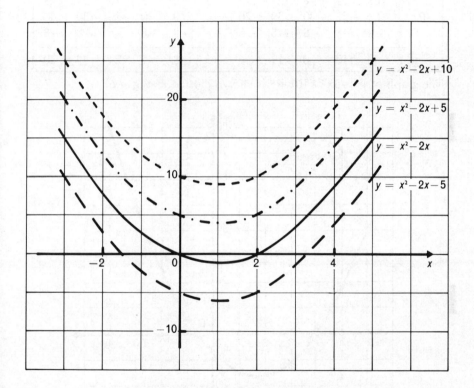

$y = x^2 - 2x + 10$

$y = x^2 - 2x + 5$

$y = x^2 - 2x$

$y = x^2 - 2x - 5$

Fig. 7.5

EXAMPLE 2

Plot the graph of $y = 3x^2 + 10x - 8$ between $x = -6$ and $x = +4$. Hence solve the equation $3x^2 + 10x - 8 = 0$.

A table can be drawn up as follows giving values of y for the chosen values of x.

x	-6	-5	-4	-3	-2	-1	0	1	2	3	4
$3x^2$	108	75	48	27	12	3	0	3	12	27	48
$10x$	-60	-50	-40	-30	-20	-10	0	10	20	30	40
-8	-8	-8	-8	-8	-8	-8	-8	-8	-8	-8	-8
y	40	17	0	-11	-16	-15	-8	5	24	49	80

The graph of $y = 3x^2 + 10x - 8$ is shown plotted in Fig. 7.6.

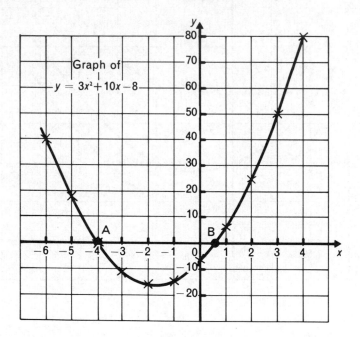

Fig. 7.6

To solve the equation $3x^2 + 10x - 8 = 0$ we have to find the value of x when $y = 0$, that is, the value of x where the graph cuts the x-axis.

These are the points A and B in Fig. 7.6.

Hence the solutions of $3x^2 + 10x - 8 = 0$ are:

$$x = -4 \quad \text{and} \quad x = 0.7$$

The accuracy of the results obtained by this method will depend on the scales chosen. In this example it would be possible to obtain a result more accurate than $x = 0.7$ by drawing the graph between the values of $x = 0$ and $x = 1$ to a much larger scale. (This has been done in Example 3 in this chapter.)

It should be noted that the value $x = -4$ is exact as this value gave $y = 0$ when selected for the table of values prior to drawing the graph.

An alternative method of solving the equation $3x^2 + 10x - 8 = 0$ is to rearrange the equation to the form $3x^2 = 8 - 10x$.

If we now plot the graphs of $y = 3x^2$ and of $y = 8 - 10x$ on the same axes, the solutions we require will be given by the values of x where the two graphs intersect.

The two graphs are shown plotted in Fig. 7.7.

The points of intersection are P and Q and the corresponding values of x are -4 and 0.7.

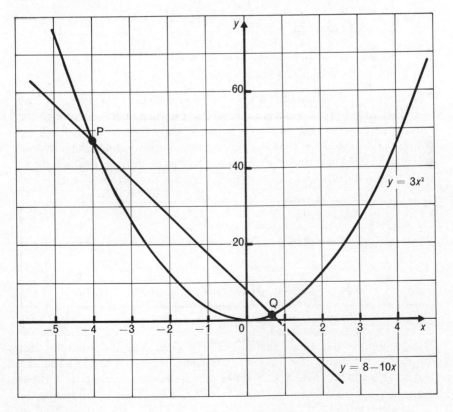

Fig. 7.7

EXAMPLE 3

Find the roots of the equation $x^2 - 1.3 = 0$ by drawing a suitable graph.

Using the same method as in Example 2 we need to plot a graph of $y = x^2 - 1.3$ and find where it cuts the x-axis.

We have not been given a range of values of x between which the curve should be plotted and so we must make our own choice.

A good method is to try first a range from $x = -4$ to $x = +4$. If only five values of y are calculated for values of x of -4, -2, 0, $+1$ and $+2$ we shall not have wasted much time if these values are not required — in any case we shall learn from this trial and be able to make a better choice at the next attempt.

The first table of values is as follows:

x	-4	-2	0	2	4
x^2	16	4	0	4	16
-1.3	-1.3	-1.3	-1.3	-1.3	-1.3
y	14.7	2.7	-1.3	2.7	14.7

The graph of these values is shown plotted in Fig. 7.8.

The approximate values of x where the curve cuts the x-axis are -1 and $+1$ (Fig. 7.8). For more accurate results we must plot the portion of the curve where it cuts the x-axis to a larger scale. We can see, however, from the table of values that the graph is symmetrical about the y-axis — so we need only plot one half. We will choose the portion to the right of the y-axis and draw up a table of values from $x = 0.7$ to $x = 1.3$:

x	0.7	0.8	0.9	1.0	1.1	1.2	1.3
x^2	0.49	0.64	0.81	1.00	1.21	1.44	1.69
-1.3	-1.3	-1.3	-1.3	-1.3	-1.3	-1.3	-1.3
y	-0.81	-0.66	-0.49	-0.30	-0.09	0.14	0.39

These values are shown plotted in Fig. 7.9.

The graph cuts the x-axis where $x = 1.14$ (Fig. 7.9) and we must not forget the other value of x where the curve cuts the x-axis to the left of the y-axis. This will be where $x = -1.14$ since the curve is symmetrical about the y-axis.

Hence the solutions of $x^2 - 1.3 = 0$ are $x = 1.14$ and $x = -1.14$.

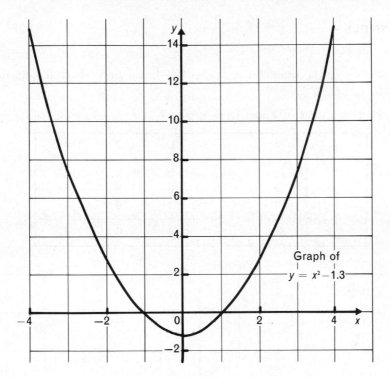

Fig. 7.8

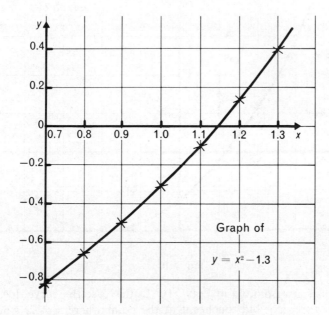

Fig. 7.9

EXAMPLE 4

Solve the equation $x^2-4x+4 = 0$.

We shall plot the graph of $y = x^2-4x+4$ and find where it cuts the x-axis.

Drawing up a table of values from $x = -4$ to $x = +4$ then:

x	-4	-2	0	2	4
x^2	16	4	0	4	16
$-4x$	16	8	0	-8	-16
$+4$	4	4	4	4	4
y	36	16	4	0	4

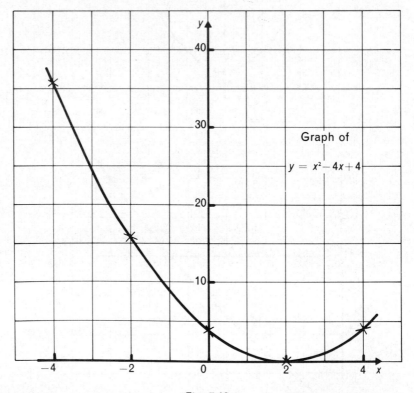

Fig. 7.10

The graph is shown plotted in Fig. 7.10. In this case the curve does not actually cut the x-axis but touches it at the point where $x = 2$. Another way of looking at it is to say that the curve 'cuts' the x-axis at two points which lie on top of each other, that is, coincide — the points are said to be coincident and give rise to repeated roots as met already in Example 9, Chapter 2.

The solution to $x^2-4x+4=0$ is, therefore, $x=2$.

EXAMPLE 5

Solve the equation $x^2+x+3=0$.

We shall plot the graph of $y=x^2+x+3$ and find where it cuts the x-axis.

Drawing up a table of values from $x=-4$ to $x=+4$. then:

x	-4	-2	0	2	4
x^2	16	4	0	4	16
x	-4	-2	0	2	4
3	3	3	3	3	3
y	15	5	3	9	23

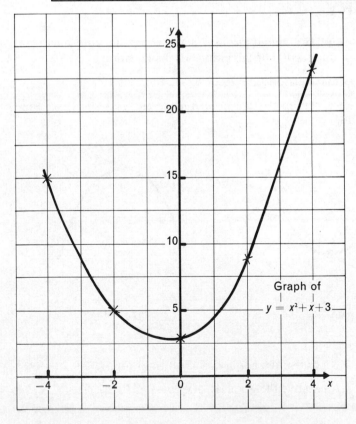

Graph of

$y = x^2 + x + 3$

Fig. 7.11

The graph is shown plotted in Fig. 7.11. We can see that the curve does not cut the x-axis at all. This means there are no roots — in theory there are roots but they are complex or imaginary and have no arithmetical value (see also Example 14, Chapter 2).

Exercise 13

By plotting suitable graphs solve the following equations:

1) $x^2 - 7x + 12 = 0$ (plot between $x = 0$ and $x = 6$)

2) $x^2 + 16 = 8x$ (plot between $x = 1$ and $x = 7$)

3) $x^2 - 9 = 0$ (plot between $x = -4$ and $x = 4$)

4) $x^2 + 2x - 15 = 0$ **7)** $x^2 - 2x - 1 = 0$

5) $3x^2 - 23x + 14 = 0$ **8)** $3x^2 - 7x + 1 = 0$

6) $2x^2 + 13x + 15 = 0$ **9)** $9x^2 - 5 = 0$

SUMMARY

a) The solutions of two simultaneous equations are the co-ordinates of the points of intersection of the graphs of the equations.

b) The graph of the equation $y = ax^2 + bx + c$ is called a parabola.

c) The vertex of a parabola is the point on the curve as shown in Figs. 7.12 and 7.13.

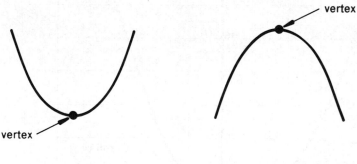

vertex

vertex

Fig. 7.12   Fig. 7.13

d) If constant a is positive the graph is shaped as in Fig. 7.12.

e) If constant a is negative the graph is shaped as in Fig. 7.13.

f) If constant b is positive the vertex is to the left of the origin.

g) If constant b is negative the vertex is to the right of the origin.

h) The vertex moves up or down according to the magnitude of constant c.

i) The solutions of the quadratic equation $ax^2+bx+c = 0$ are the values of x where the curve cuts the horizontal x-axis.

j) Fig. 7.14 shows the graph actually cutting the x-axis. This gives two solutions (real roots).

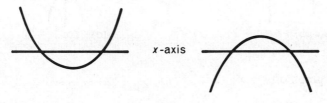

Fig. 7.14

k) Fig. 7.15 shows the graph tangential to the x-axis. This gives one solution (repeated roots).

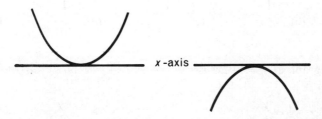

Fig. 7.15

l) Fig. 7.16 shows the graph not cutting the x-axis at all. This gives no real solutions (complex roots).

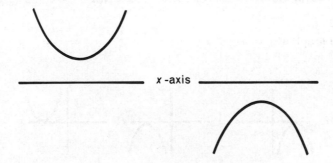

Fig. 7.16

Self-Test 3

State which answer or answers are correct.

In every diagram the origin is at the intersection of the axes.

1) The graph of $y = x^2$ is:

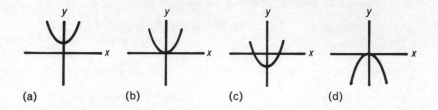

(a)   (b)   (c)   (d)

2) The graph of $y = x^2 + 2$ is:

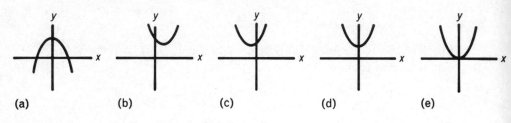

(a)   (b)   (c)   (d)   (e)

3) The graph of $y = -2x^2$ is:

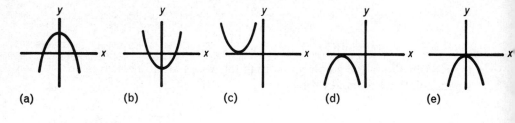

(a)   (b)   (c)   (d)   (e)

4) The graph of $y = 4 - x^2$ is:

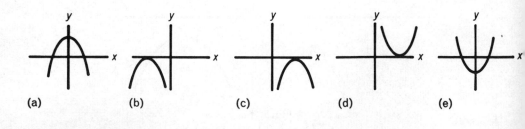

(a)   (b)   (c)   (d)   (e)

5) The graph of $y = x^2 - 3x + 2$ is:

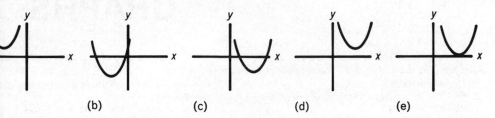

(b)   (c)   (d)   (e)

6) The graph of $y = x^2 + 4x + 4$ is:

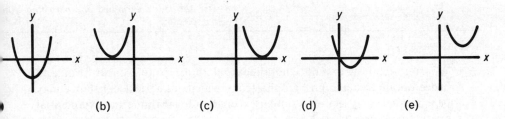

(b)   (c)   (d)   (e)

LOGARITHMIC GRAPHS

We have seen how to reduce non-linear relationships to a linear form, and hence obtain straight line graphs from which numerical constants may be evaluated. This section is similar except that logarithmic and exponential relationships are covered.

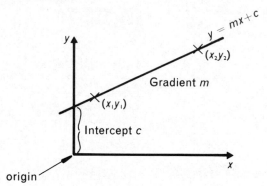

Fig. 8.1

Remember the fundamental equation of the straight line:

$$y = mx + c$$

If m and c cannot be found directly from inspection of the graph, two points (x_1, y_1) and (x_2, y_2) on the line are chosen (Fig. 8.1) and their values substituted in the linear equation, giving the simultaneous equations:

$$y_1 = mx_1 + c$$
$$y_2 = mx_2 + c$$

from which m and c can be found.

Consider the following relationships, in which z and t are the variables whilst a, b and n are constants.

1)
$$z = a \cdot t^n$$

Taking logs we get:

$$\log z = \log(t^n \cdot a)$$
$$= \log t^n + \log a$$
$$\therefore \quad \log z = n \cdot (\log t) + \log a$$

Comparing this with the equation $y = mx + c$ of a straight line, we see that if we plot $\log z$ on the y-axis and $\log t$ on the x-axis (Fig. 8.2.) then the resulting straight line will have gradient equal to n and an intercept on the vertical axis equal to $\log a$.

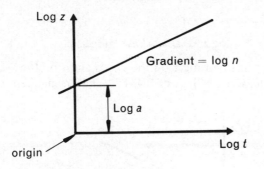

Fig. 8.2

2)
$$z = a \cdot b^t$$

Taking logs we get:

$$\log z = \log(b^t \cdot a)$$
$$= \log b^t + \log a$$
$$\therefore \quad \log z = (\log b) \, t + \log a$$

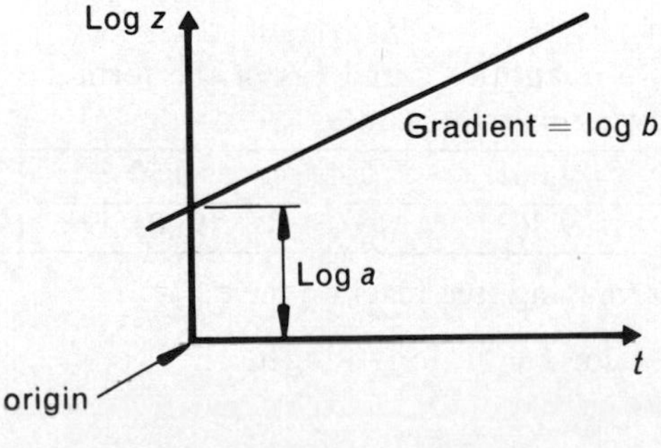

Fig. 8.3

To obtain a straight line plot log z on the y-axis and t on the x-axis (Fig. 8.3). The line will have a gradient equal to log b and an intercept on the y-axis equal to log a.

3) $$z = a \cdot e^{bt}$$

where e is the base of Naperian logarithms and has a value of 2.718 3 correct to four places of decimals.

Taking logs we get:

$$\log z = \log (e^{bt} \cdot a)$$
$$= \log e^{bt} + \log a$$
$$= bt \cdot (\log e) + \log a$$
$$\therefore \qquad \log z = (b \cdot \log e) t + \log a$$

But log e $= 0.434\ 3$,

$$\therefore \qquad \log z = (0.434\ 3b) t + \log a$$

To obtain a straight line plot log z on the y-axis and t on the x-axis (Fig. 8.4). The line will have a gradient of 0.434 3b and an intercept on y-axis equal to log a.

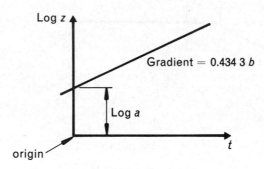

Fig. 8.4

EXAMPLE 1

The law connecting two quantities z and t is of the form $z = a.t^n$. Find the law given the following pairs of values:

z	3.170	4.603	7.499	10.50	15.17
t	7.980	9.863	13.03	15.81	19.50

By taking logs and rearranging (see text) we have:

$$\log z = n \cdot \log t + \log a \qquad (1)$$

From the given values:

log z	0.501 1	0.663 1	0.875 0	1.021 2	1.181 0
log t	0.902 0	0.994 0	1.114 9	1.199 0	1.290 0

Since it is not convenient to show the origin (point 0, 0) we shall use the two-point method of finding the constants (Fig. 8.5).

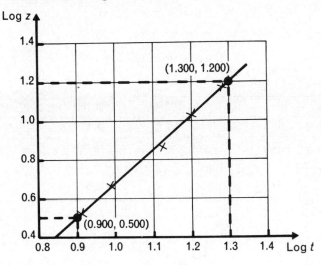

Fig. 8.5

Point (0.900, 0.500) lies on the line, and substituting in equation (1),

$$0.500 = n(0.900) + \log a \qquad (2)$$

Point (1.300, 1.200) lies on the line, and substituting in equation (1),

$$1.200 = n(1.300) + \log a \qquad (3)$$

Subtracting equation (2) from equation (3),

$$0.7 = 0.4n$$

∴
$$n = 1.75$$

Substituting in the equation (2),

$$0.500 = 1.75(0.900) + \log a$$

$$\log a = -1.075$$

$$= -2 + 0.925$$

$$= \bar{2}.925$$

$$a = 0.084$$

EXAMPLE 2

In an experiment the following values were obtained:

t	0.190	0.250	0.300	0.400
z	11 220	18 620	26 920	61 660

It is thought that the law may be of the form $z = a.b^t$ where a and b are constants. Verify this and find the law.

By taking logs and rearranging, we have:

$$\log z = (\log b)\, t + \log a \qquad (1)$$

From the given values:

t	0.190	0.250	0.300	0.400
$\log z$	4.050	4.270	4.430	4.790

The plotted points lie in a straight line (Fig. 8.6), We have verified the straight line relationship $\log z = (\log b)\, t + \log a$, and hence $z = a.b^t$

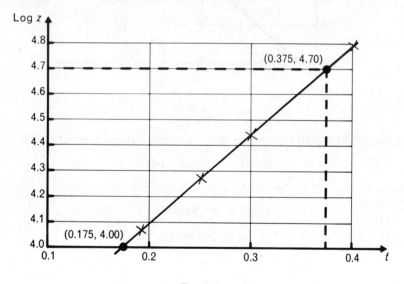

Fig. 8.6

As in the last example we shall use the two-point method of finding the constants.

Point (0.375, 4.70) lies on the line, and substituting in equation (1),

$$4.7 = (\log b)0.375 + \log a \qquad (2)$$

Point (0.175, 4.00) lies on the line, and substituting in equation (1),

$$4.0 = (\log b)0.175 + \log a \qquad (3)$$

Subtracting equation (3) from equation (2),

$$0.7 = (\log b)0.2$$

$$\therefore \qquad \log b = 3.5$$

$$\therefore \qquad b = 3162$$

Substituting in the equation (2),

$$4.7 = 3.5 \times 0.375 + \log a$$

$$\therefore \qquad \log a = 4.7 - 1.312\,5$$

$$= 3.387\,5$$

$$\therefore \qquad a = 2441$$

Hence the required law is:

$$z = 2441(3162)^t$$

EXAMPLE 3

V and t are connected by the law $V = a.e^{bt}$. We are given the following tables of values. Establish values for the constants a and b:

t	0.05	0.95	2.05	2.95
V	20.70	24.49	30.27	36.06

By taking logs and rearranging we have:

$$\log V = (0.434\,3b)\,t + \log a \qquad (1)$$

From the given values:

t	0.05	0.95	2.05	2.95
$\log V$	1.316	1.389	1.481	1.557

Once again the two-point method (Fig. 8.7) is used to find the constants:

Point (3.25, 1.579) lies on the line and substituting in equation (1),

$$\therefore \qquad 1.579 = (0.434\,3b)3.25 + \log a \qquad (2)$$

Point (0.25, 1.330) lies on the line, and substituting in equation (1),

$$\therefore \qquad 1.330 = (0.434\,3b)0.25 + \log a \qquad (3)$$

Subtracting equation (3) from equation (2),

$$0.249 = (0.434\,3 \times 3)b$$

$$b = 0.191$$

Substituting in the equation (2),

$$1.330 = (0.434\,3)0.191(0.25) + \log a$$

$$\therefore \qquad \log a = 1.309\,3$$

$$a = 20.4$$

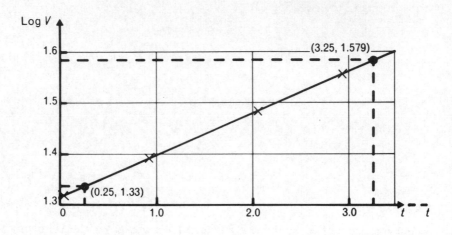

Fig. 8.7

Exercise 14

1) The following values of x and y follow a law of the type $y = ax^n$. By plotting $\log y$ (vertically) against $\log x$ (horizontally) find values for a and n.

x	1	2	3	4	5
y	3	12	27	48	75

2) The following results were obtained in an experiment to find the relationship between the luminosity I of a metal filament lamp and the voltage V.

V	40	60	80	100	120
I	5.1	26.0	82	200	414

The law is thought to be of the type $I = aV^n$. Test this by plotting $\log I$ (vertically) against $\log V$ (horizontally) and find suitable values for a and n.

3) The relationship between power P (watts), the e.m.f. E (volts) and the resistance R (ohms) is thought to be of the form $P = \dfrac{E^n}{R}$. In an experiment in which R was kept constant the following results were obtained:

E	5	10	15	20	25	30
P	2.5	10	22.5	40	62.5	90

Verify the law and find the values of the constants n and R.

4) The following values were obtained experimentally:

t	0.25	0.50	1.00	3.00	5.00	7.00
E	7.74	6.51	4.60	1.15	0.29	0.07

The relationship connecting E and t is thought to be of the form $E = a.b^t$ where a and b are constants. Verify this and find the law.

5) For a constant pressure process on a certain gas the formula connecting the absolute temperature T and the specific entropy s is of the form $T = ke^{cs}$ where e is the logarithmic base and k and c are constants. When $T = 460$, $s = 1.000$, and when $T = 600$, $s = 1.089$. Find constants k and c to three significant figures.

6) The instantaneous e.m.f., v, induced in a coil after a time, t, is given by $v = V.e^{-t/T}$ where V and T are constants. Find the values of V and T given the following values:

v	95	80	65	40	25
t	0.000 13	0.000 56	0.001 08	0.002 29	0.003 47

LOGARITHMIC SCALES

The examples given earlier in this chapter which involved logarithmic and exponential laws were solved using ordinary graph paper. This entailed finding the logs of each individual given number on at least one of the axes. A slide rule has a logarithmic scale, and if this is used to mark off the divisions where log values are to be plotted it is no longer necessary to find individual logs.

An example of a logarithmic scale constructed by using a slide rule is shown in Fig. 8.8.

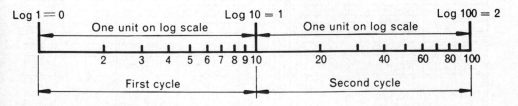

Fig. 8.8

It may be seen that the scale comprises repeated patterns called cycles. The choice of numbers on the scale depends on the numbers allocated to the variable in the problem to be solved — thus in Fig. 8.8 the numbers run from 1 to 100.

Other examples of log scales are shown in Figs. 8.9 and 8.10.

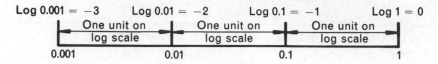

Fig. 8.9

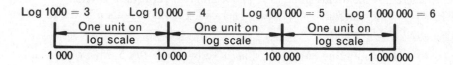

Fig. 8.10

LOGARITHMIC GRAPH PAPER

Logarithmic scales (such as those used on a slide rule) may be used on graph paper in place of the more usual linear scales. By using graph paper ruled in this way log plots may be made without the necessity of looking up the logs of each given value. Semi-logarithmic graph paper is also available and has one way ruled with log scales whilst the other way has the usual linear scale. Examples of each are shown in Fig. 8.11.

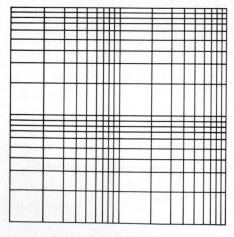

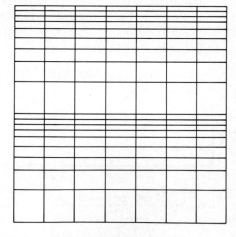

Full logarithmic rulings   Semi-logarithmic rulings

Fig. 8.11

The use of logarithmic graph paper enables power or exponential relationships between two variables to be verified quickly.

A straight line graph on full logarithmic (or log-log) graph paper indicates a relationship between the variables x and y of the form $y = a.x^n$, a and n being constants.

A straight line graph on semi-logarithmic (or log-linear) graph paper indicates a relationship between the variables x and y of the form $y = a.b^x$, a and b being constants.

The use of these scales and the special graph paper is shown by solving the three problems already completed using ordinary linear paper.

EXAMPLE 4

The law connecting two quantities z and t is of the form $z = a.t^n$. Find the law given the following pairs of values:

z	3.170	4.603	7.499	10.50	15.17
t	7.980	9.863	13.03	15.81	19.50

The relationship,

$$z = a \cdot t^n$$

gives, by taking logs of both sides,

$$\log z = n.\log t + \log a \qquad (1)$$

Instead of looking up each $\log z$ and each $\log t$ individually we can plot the given values of z and t on log scales as shown in Fig. 8.12.

Choice of scales is largely governed by the log graph paper, the scales comprising a repeating pattern. Each cycle is ten times the previous one (e.g. 0.1 to 1, 1 to 10, 10 to 100, etc.).

The constants are again found by taking two pairs of co-ordinates:

Point (25, 23) lies on the line and putting these values in equation (1),

$$\log 23 = n(\log 25) + \log a \qquad (2)$$

Point (4.1, 1) lies on the line and putting these values in equation (1),

$$\log 1 = n(\log 4.1) + \log a \qquad (3)$$

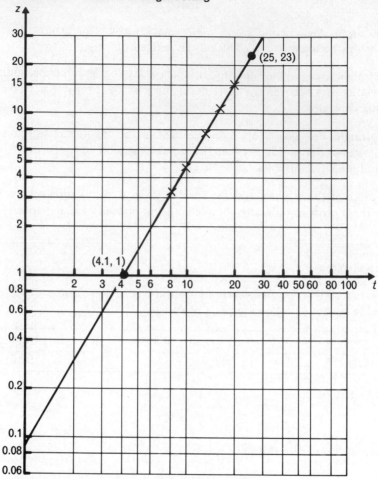

Fig. 8.12

Subtracting equation (3) from equation (2),

$$\log 23 - \log 1 = n(\log 25 - \log 4.1)$$

$$\therefore \qquad \log\left(\frac{23}{1}\right) = n\{\log\left(\frac{25}{4.1}\right)\}$$

$$\therefore \qquad n = \frac{\log(23/1)}{\log(25/4.1)} = \frac{\log 23}{\log 6.1} = \frac{1.361\,7}{0.785\,3}$$

$$\therefore \qquad n = 1.74$$

Substituting this value of n in equation (3),

$$\log 1 = 1.74(\log 4.1) + \log a$$

$$\log a = \log 1 - 1.74(\log 4.1)$$
$$= 0 - 1.075$$
$$= -1.075 = -1-1+1-0.075$$
$$= -2+0.925 = \bar{2}.925$$
$$\therefore \qquad a = 0.084$$

Hence the law is $z = 0.084t^{1.74}$.

These results verify those obtained by the other method using ordinary graph paper (see Example 1).

EXAMPLE 5

The table gives values obtained in an experiment. It is thought that the law may be of the form $z = a.b^t$, where a and b are constants. Verify this and find the law.

t	0.190	0.250	0.300	0.400
z	11 220	18 620	26 920	61 660

We think that the relationship is of the form:

$$z = a \cdot b^t$$

which gives, by taking logs of both sides,

$$\log z = (\log b)\, t + \log a \qquad (1)$$

Instead of looking up logs corresponding to each value of z we shall plot the given values on a vertical log scale — the t values however will be on the horizontal axis using an ordinary linear scale (Fig. 8.13).

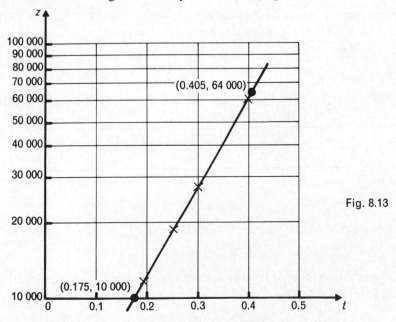

Fig. 8.13

The points lie on a straight line, and as before the values obey the law. We now have to find the co-ordinates of two points lying on the line.

Point (0.405, 64 000) lies of the line, and substituting in equation (1),

$$\log 64\,000 = (\log b)0.405 + \log a \qquad (2)$$

Point (0.175, 10 000) lies on the line, and substituting in equation (1),

$$\log 10\,000 = (\log b)0.175 + \log a \qquad (3)$$

Subtracting equation (3) from equation (2),

$$\log 64\,000 - \log 10\,000 = (\log b)\,(0.405 - 0.175)$$

$$\therefore \qquad \log b = \frac{\log (64\,000/10\,000)}{0.230}$$

$$\therefore \qquad b = 3162$$

Substituting in the equation (3),

$$\log 10\,000 = (3.5)0.175 + \log a$$

$$\therefore \qquad \log a = \log 10\,000 - 3.5(0.175)$$

$$= 4 - 0.613 = 3.387$$

$$\therefore \qquad a = 2441$$

Hence the law is:

$$z = 2441(3162)^t$$

which verifies the previous result given in Example 2.

EXAMPLE 6

V and t are connected by the law $V = a.e^{bt}$. If the values given in the table satisfy the law, find the constants a and b.

t	0.05	0.95	2.05	2.95
V	20.70	24.49	30.27	36.06

The law is:

$$V = a \cdot e^{bt}$$

which gives, by taking logs of both sides,

$$\log V = 0.434\,3\,bt + \log a \qquad (1)$$

As in the last example V values are plotted on a log scale on the vertical axis, whilst the t values are plotted on the horizontal axis on an ordinary linear scale (Fig. 8.14).

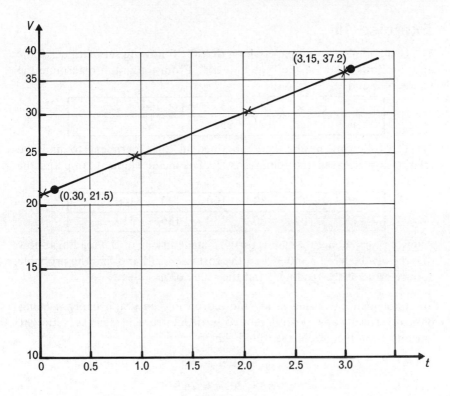

Fig. 8.14

Point (3.15, 37.2) lies on the line, and substituting in equation (1),

$$\log 37.2 = (0.434\,3)b(3.15) + \log a \qquad (2)$$

Point (0.30, 21.5) lies on the line, and substituting in equation (1),

$$\log 21.5 = (0.434\,3)b(0.3) + \log a \qquad (3)$$

Subtracting,

$$\log 37.2 - \log 21.5 = (0.434\,3)b(3.15 - 0.3)$$

$$\therefore \qquad b = \frac{\log(37.2/21.5)}{(0.434\,3)(2.85)} = \frac{0.238}{(0.434\,3)(2.85)}$$

$$= 0.192$$

Substituting in equation (3),

$$\log 21.5 = (0.434\,3)0.192(0.3) + \log a$$

$$\therefore \qquad \log a = 1.332\,4 - 0.024\,8 = 1.307\,6$$

$$a = 20.3$$

Exercise 15

1) Using log-log graph paper show that the following set of values for x and y follows a law of the type $y = ax^n$. From the graph determine the values of a and n.

x	4	16	25	64	144	296
y	6	12	15	24	36	52

2) The following results were obtained in an experiment to find the relationship between the luminosity I of a metal filament lamp and the voltage V:

V	60	80	100	120	140
I	11	20.5	89	186	319

Allowing for the fact than an error was made in one of the readings show that the law between I and V is of the form $I = aV^n$ and find the probable correct value of the reading. Find the value of n.

3) Two quantities t and m are plotted on log-log graph paper, t being plotted vertically and m being plotted horizontally. The result is a straight line and from the graph it is found that:

when $m = 8, \quad t = 6.8$

and when $m = 20, t = 26.9$

Find the law connecting t and m.

4) The intensity of radiation, R, from certain radioactive materials at a particular time t is thought to follow the law $R = kt^n$. In an experiment to test this the following values were obtained:

R	58	43.5	26.5	14.5.	10
t	1.5	2	3	5	7

Show that the assumption was correct and evaluate k and n.

5) The values given in the following table are thought to obey a law of the type $y = ab^{-x}$. Check this statement and find the values of the constants a and b.

x	0.1	0.2	0.4	0.6	1.0	1.5	2.0
y	175	158	60	32	6.4	1.28	0.213

6) The force F on the tight side of a driving belt was measured for different values of the angle of lap θ and the following results were obtained:

F	7.4	11.0	17.5	24.0	36.0
θ rad	$\pi/4$	$\pi/2$	$3\pi/4$	π	$5\pi/4$

Construct a graph to show these values conform approximately to an equation of the form $F = ke^{\mu\theta}$. Hence find the constants μ and k.

7)   A capacitor and resistor are connected in series. The current i amperes after time t seconds is thought to be given by the equation $i = I.e^{-t/T}$ where I amperes is the initial charging current and T seconds is the time constant. Using the following values verify the relationship and find the values of the constants I and T:

i amperes	0.015 6	0.012 1	0.019 45	0.007 36	0.005 73
t seconds	0.05	0.10	0.15	0.20	0.25

SUMMARY

a) To verify the relationship $z = a.t^n$

rewrite in the form $\log z = n(\log t) + \log a$

and then plot $\log z$ against $\log t$ in order to obtain a straight line graph.

b) To verify the relationship $z = a.b^t$

rewrite in the form $\log z = (\log b)t + \log a$

and then plot $\log z$ against t in order to obtain a straight line graph.

c) To verify the relationship $z = a.e^{bt}$

rewrite in the form $\log z = (0.434\ 3b)t + \log a$

and then plot $\log z$ against t in order to obtain a straight line graph.

EXPONENTIAL GRAPHS

VALUES OF e^x AND e^{-x}

Most mathematical tables which include logarithms and trignometrical functions also include a table giving values of e^x and e^{-x} for values of x from 0 to 6. It may appear that the range of x values is rather limited but it is adequate for most practical problems as may be seen from the examples which follow later in this chapter.

EXPONENTIAL GRAPHS

Curves which have equations of the type e^x and e^{-x} are called exponential graphs.

We may plot the graphs of e^x and e^{-x} by using mathematical tables to find values of e^x and e^{-x} for chosen values of x. We should remember that any number to a zero power is unity: hence $e^0 = 1$.

Drawing up a table of values we have:

x	-2	-1	0	1	2
e^x	0.14	0.37	1	2.72	7.39
$-x$	2	1	0	-1	-2
e^{-x}	7.39	2.72	1	0.37	0.14

For convenience both the curves are shown plotted on the same axes in Fig. 9.1. Although the range of values chosen for x is limited the overall shape of the curves is clearly shown.

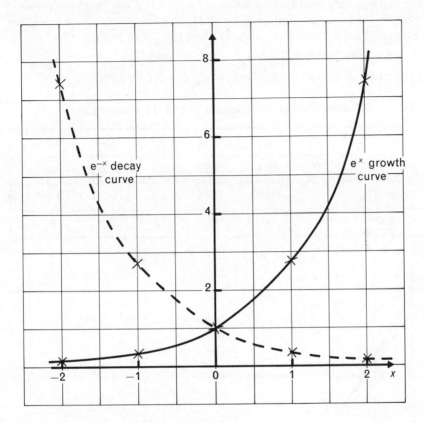

Fig. 9.1

The rate at which a curve is changing at any point is given by the gradient of the tangent at that point.

Remember the sign convention for gradients is:

positive
gradient

negative
gradient

Now the gradient at any point on the e^x graph is positive and so the rate of change is positive. In addition the rate of change increases as the values of x increase. A graph of this type is called a growth curve.

The gradient at any point on the e^{-x} graph is negative and so the rate of change is negative. In addition the rate of change decreases as the values of x decrease. A graph of this type is called a decay curve.

EXAMPLE 1

The instantaneous e.m.f. in an inductive circuit is given by the expression $100.e^{-4t}$ volts, where t is time in seconds. Plot the graph of the e.m.f. for values of t from 0 to 0.5 seconds, and use the graph to find:

(a) the value of the e.m.f. when $t = 0.25$ seconds and

(b) the rate of change of the e.m.f. when $t = 0.1$ seconds.

First we will draw up a table of values of t at 0.1 s intervals:

t	0	0.1	0.2	0.3	0.4	0.5
$-4t$	0	-0.4	-0.8	-1.2	-1.6	-2.0
e^{-4t}	1	0.67	0.45	0.30	0.20	0.14
$100e^{-4t}$	100	67	45	30	20	14

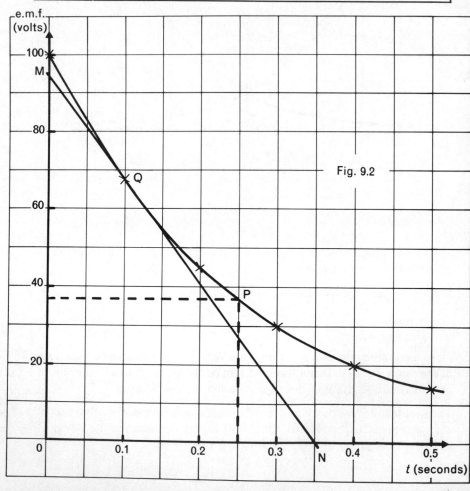

Fig. 9.2

The graph is shown plotted in Fig. 9.2.

(a) The point P on the curve is at 0.25 seconds shown on the t scale and the corresponding value of e.m.f. can be read directly from the vertical axis scale. The value is 37 volts.

(b) The point Q on the graph is at 0.1 seconds. Now the rate of change of the curve at Q is given by the gradient of the tangent at Q. This gradient may be found by constructing a suitable right angled triangle such as MNO in Fig. 9.2, and finding the ratio $\dfrac{MO}{ON}$.

Hence the gradient at Q $= \dfrac{MO}{ON} = \dfrac{94 \text{ volts}}{0.35 \text{ seconds}} = 269$ volts per second.

According to the sign convention a line sloping downwards from left to right has a negative gradient.

Hence the gradient at Q is -270 volts per second, which means that the rate of change of the curve at Q is -270 volts per second.

This is the same as saying that the e.m.f. at $t = 0.1$ seconds is decreasing at the rate of 270 volts per second.

EXAMPLE 2

The formula $i = 2(1 - e^{-10t})$ gives the relationship between the instantaneous current i amperes and the time t seconds in an inductive circuit. Plot a graph of i against t taking values of t from 0 to 0.3 seconds at intervals of 0.05 seconds. Hence find:

(a) the initial rate of growth of the current i when $t = 0$, and

(b) the time taken for the current to increase from 1 to 1.6 amperes.

The table of values is drawn up as follows:

t	0	0.05	0.10	0.15	0.20	0.25	0.30
$-10t$	0	-0.5	-1.0	-1.5	-2.0	-2.5	-3.0
1	1	1	1	1	1	1	1
e^{-10t}	1	0.61	0.37	0.22	0.14	0.08	0.05
$1 - e^{-10t}$	0	0.39	0.63	0.78	0.86	0.92	0.95
$2(1 - e^{-10t})$	0	0.78	1.26	1.56	1.72	1.84	1.90

The curve is shown plotted in Fig. 9.3.

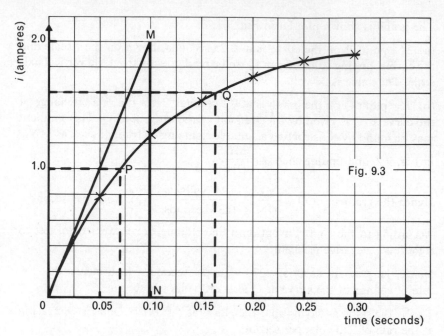

Fig. 9.3

(a) When $t = 0$ the initial rate of growth will be given by the gradient of the tangent at O. The tangent at O is the line OM and its gradient may be found by using a suitable right angled triangle MNO and finding the ratio $\dfrac{MN}{ON}$.

Hence the initial rate of growth of $\quad i = \dfrac{MN}{ON} = \dfrac{2 \text{ amperes}}{0.1 \text{ seconds}}$

$$= 20 \text{ amperes per second.}$$

(b) The point P on the curve corresponds to a current of 1.0 amperes and the time at which this occurs may be read from the t scale and is 0.07 seconds.

Similarly point Q corresponds to a 1.6 ampere current and occurs at 0.16 seconds.

Hence the time between P and Q is $0.16 - 0.07 = 0.09$ seconds.

This means that the time for the current to increase from 1 to 1.6 amperes is 0.09 seconds.

Exercise 16

1) Use tables of exponential functions to find the values of:

(a) $e^{0.3}$ (b) $e^{2.5}$ (c) $e^{4.5}$

(d) $e^{-0.4}$ (e) e^{-1} (f) $e^{-3.5}$

2) Plot a graph of $y = e^{2x}$ for values of x from -1 to $+1$ at 0.25 unit intervals. Use the graph to find the value of y when $x = 0.3$, and the value of x when $y = 5.4$.

3) Using values of x from -4 to $+4$ at one unit intervals plot a graph of $y = e^{-x/2}$. Hence find the value of x when $y = 2$, and the gradient of the curve when $x = 0$.

4) For a constant pressure process on a certain gas the formula connecting the absolute temperature T and the specific entropy s is $T = 24.e^{3s}$. Plot a graph of T against s taking values of s equal to 1.000, 1.033, 1.066, 1.100, 1.133, 1.166 and 1.200. Use the graph to find the value of

(a) T when $s = 1.09$.

(b) s when $T = 700$.

5) The equation $i = 2.4e^{-6t}$ gives the relationship between the instantaneous current, i mA, and the time, t seconds.
Plot a graph of i against t for values of t from 0 to 0.6 seconds at 0.1 second intervals.
Use the curve obtained to find the rate at which the current is decreasing when $t = 0.2$ seconds.

6) In a capacitive circuit the voltage v and the time t seconds are connected by the relationship $v = 240(1 - e^{-5t})$.
Draw the curve of v against t for values of $t = 0$ to $t = 0.7$ seconds at 0.1 second intervals.
Hence find:
(a) the time when the voltage is 140 volts, and,
(b) the initial rate of growth of the voltage when $t = 0$.

GRADIENT OF A CURVE

GRAPHICAL METHOD

In mathematics and engineering we often need to know the rate of change of one variable with respect to another. For instance, speed is the rate of change of distance with respect to time, and acceleration is the rate of change of velocity with respect to time.

Consider the graph of $y = x^2$, part of which is shown in Fig. 10.1. As the values of x increase so do the values of y, but they do not increase at the same rate. A glance at the portion of the curve shown shows that the values of y increase faster when x is large, because the gradient of the curve is increasing.

To find the rate of change of y with respect to x at a particular point we need to find the gradient of the curve at that point.

If we draw a tangent to the curve at the point the gradient of the tangent will be the same as the gradient of the curve.

Fig. 10.1

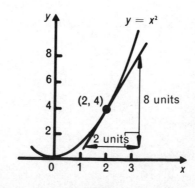

EXAMPLE 1

Find the gradient of the curve $y = x^2$ at the point where $x = 2$.

The point where $x = 2$ is the point $(2, 4)$. We draw a tangent at this point, as shown in Fig. 10.1. Then by constructing a right-angled triangle the gradient is found to be $\dfrac{8}{2} = 4$. This gradient is positive, in accordance with our previous work, since the tangent slopes upwards from left to right.

EXAMPLE 2

Find the gradient of the curve $y = x^2$ at the point where $x = -2$.

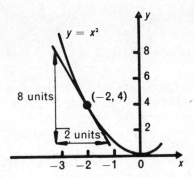

Fig. 10.2

The point where $x = -2$ is the point $(-2, 4)$. By drawing a tangent here, as in Fig. 10.2, and constructing a right-angled triangle as before, the gradient is found to be $\dfrac{-8}{2} = -4$. This gradient is negative since the tangent slopes downwards from left to right.

EXAMPLE 3

Draw the graph of $y = x^2 - 3x + 7$ between $x = -4$ and $x = 3$ and hence find the gradient at:
(a) the point $x = -3$,
(b) the point $x = 2$.

(a) At the point where $x = -3$,

$$y = (-3)^2 - 3(-3) + 7 = 25$$

At the point $(-3, 25)$ draw a tangent as shown in Fig. 10.3. The gradient is found by drawing a right-angled triangle (which should be as large as conveniently possible for accuracy) as shown, and measuring its height and base.

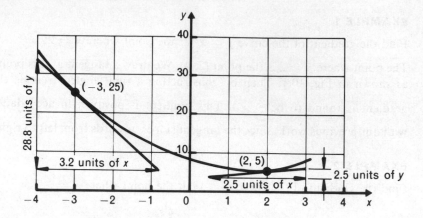

Fig. 10.3

Hence

$$\text{gradient at point } (-3, 25) = -\frac{28.8}{3.2} = -9$$

the negative sign indicating a downward slope from left to right.

(b) At the point where $x = 2$,

$$y = 2^2 - 3(2) + 7 = 5$$

Hence by drawing a tangent and a right-angled triangle at the point $(2, 5)$ in a similar manner to above,

$$\text{gradient at point } (2, 5) = \frac{2.5}{2.5} = 1$$

being positive as the tangent slopes upwards from left to right.

Exercise 17

1) Find the gradient of the curve $y = 3x^2 + 7x + 3$ at the points where $x = -2$ and $x = 2$.

2) Find the gradient of the curve $y = 2x^3 - 7x^2 + 5x - 3$ at the points where $x = -1.5$, $x = 0$ and $x = 3$.

3) Draw the graph of $2x^2 - 5$ for values of x between -2 and $+3$. Draw, as accurately as possible, the tangents to the curve at the points where $x = -1$ and $x = +2$ and hence find the gradient of the curve at these points.

4) Draw the curve $y = x^2 - 3x + 2$ from $x = 2.5$ to $x = 3.5$ and find its gradient at the point where $x = 3$.

5) For what values of x is the gradient of the curve

$$y = \frac{x^3}{3} + \frac{x^2}{2} - 33x + 7$$

equal to 3?

6) For what values of x is the gradient of the curve $y = 3 + 4x - x^2$ equal to:

(a) -1,   (b) 0,   (c) 2?

7) Draw the curve

$$y = x - \frac{1}{x}$$

from $x = 0.8$ to 1.2. Find its gradient at $x = 1$.

NUMERICAL METHOD

The gradient of a curve may always be found by graphical means but this method is often inconvenient. A numerical method will now be developed.

Consider the curve $y = x^2$. Part of this is shown in Fig. 10.4. Let P be the point on the curve at which $x = 1$ and $y = 1$. Q is a variable point on the curve, which will be considered to start at the point (2, 4) and move down the curve towards P, rather like a bead slides down a wire.

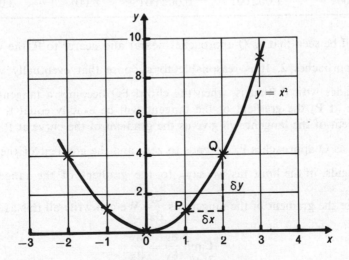

Fig. 10.4

The symbol δx will be used to represent an increment of x, and δy will be used to represent the corresponding increment of y. The gradient of the chord PQ is then

$$\frac{\delta y}{\delta x}$$

When Q is at the point (2, 4) then $\delta x = 1$, and $\delta y = 3$

$$\therefore \qquad \frac{\delta y}{\delta x} = \frac{3}{1} = 3$$

The following table shows how $\dfrac{\delta y}{\delta x}$ alters as Q moves nearer and nearer to P.

Coordinates of Q				Gradient of
x	y	δy	δx	$PQ = \dfrac{\delta y}{\delta x}$
2	4	3	1	3
1.5	2.25	1.25	0.5	2.5
1.4	1.96	0.96	0.4	2.4
1.3	1.69	0.69	0.3	2.3
1.2	1.44	0.44	0.2	2.2
1.1	1.21	0.21	0.1	2.1
1.01	1.020 1	0.020 1	0.01	2.01
1.001	1.002 001	0.002 001	0.001	2.001

It will be seen that as Q approaches nearer and nearer to P, the value of $\dfrac{\delta y}{\delta x}$ approaches 2. It is reasonable to suppose that eventually when Q coincides with P (that is, when the chord PQ becomes a tangent to the curve at P) the gradient of the tangent will be exactly equal to 2. The gradient of the tangent will give us the gradient of the curve at P.

Now as Q approaches P, δx tends to zero and the gradient of the chord, $\dfrac{\delta y}{\delta x}$, tends, in the limit (as we say), to the gradient of the tangent. We denote the gradient of the tangent as $\dfrac{dy}{dx}$. We can write all this as

$$\underset{\delta x \to 0}{\text{Limit}} \frac{\delta y}{\delta x} = \frac{dy}{dx}$$

Instead of selecting special values for δy and δx let us now consider the general case, so that P has the coordinates (x, y) and Q has the coordinates

$(x+\delta x, y+\delta y)$, (Fig. 10.5). Q is taken very close to P, so that δx is a very small quantity.

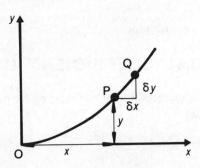

Fig. 10.5

Now:

$$y = x^2$$

and as $Q(x+\delta x, y+\delta y)$ lies on the curve, then:

$$y+\delta y = (x+\delta x)^2$$

$$\therefore \qquad y+\delta y = x^2+2x\delta x+(\delta x)^2$$

But $y = x^2$, and so we get:

$$\delta y = 2x\delta x+(\delta x)^2$$

and, by dividing both sides by δx, the gradient of chord PQ is

$$\frac{\delta y}{\delta x} = 2x+\delta x$$

As Q approaches P, δx tends to zero and $\frac{\delta y}{\delta x}$ tends, in the limit, to the gradient of the tangent of the curve at P. Thus

$$\underset{\delta x \to 0}{\text{Limit}} \frac{\delta y}{\delta x} = \frac{dy}{dx} = 2x$$

The process of finding $\frac{dy}{dx}$ is called *differentiation*.

The symbol $\frac{dy}{dx}$ means the differential coefficient of y with respect to x.

We can now check our assumption regarding the gradient of the curve at P. Since at P the value of $x = 1$, then

$$\frac{dy}{dx} = 2 \times 1 = 2$$

and we see that our assumption was correct.

DIFFERENTIAL COEFFICIENT OF x^n

It can be shown, by a method similar to that used for finding the differential coefficient of x^2, that :

If

$$y = x^n$$

then

$$\frac{dy}{dx} = nx^{n-1}$$

This is true for all values of n including negative and fractional indices. When we use it as formula it enables us to avoid having to differentiate each time from first principles.

EXAMPLES 4

If $\quad y = x^3,$

then $\quad \dfrac{dy}{dx} = 3x^2$

If $\quad y = \dfrac{1}{x} = x^{-1},$

then $\quad \dfrac{dy}{dx} = -x^{-2} = -\dfrac{1}{x^2}$

If $\quad y = \sqrt{x} = x^{1/2},$

then $\quad \dfrac{dy}{dx} = \tfrac{1}{2}x^{-1/2} = \tfrac{1}{2}\dfrac{1}{x^{1/2}} = \dfrac{1}{2\sqrt{x}}$

If $\quad y = \sqrt[5]{x^2} = x^{2/5},$

then $\quad \dfrac{dy}{dx} = \dfrac{2}{5}x^{2/5-1} = \dfrac{2}{5}x^{-3/5} = \dfrac{2}{5\sqrt[5]{x^3}}$

When a power of x is multiplied by a constant, that constant remains unchanged by the process of differentiation.

Hence if

$$y = ax^n$$

then

$$\frac{dy}{dx} = anx^{n-1}$$

EXAMPLES 5

If $y = 3x^4$,

then $\dfrac{dy}{dx} = 3 \times 4x^3 = 12x^3$

If $y = 2x^{1.3}$,

then $\dfrac{dy}{dx} = 2(1.3)x^{0.3} = 2.6x^{0.3}$

If $y = \frac{1}{5}x^7$,

then $\dfrac{dy}{dx} = \frac{1}{5} \times 7x^6 = \frac{7}{5}x^6$

If $y = \frac{3}{4} \sqrt[3]{x} = \frac{3}{4}x^{1/3}$,

then $\dfrac{dy}{dx} = \frac{3}{4} \times \frac{1}{3}x^{-2/3} = \frac{1}{4}x^{-2/3}$

If $y = \dfrac{4}{x^2} = 4x^{-2}$,

then $\dfrac{dy}{dx} = 4(-2)x^{-3} = -8x^{-3}$

When a numerical constant is differentiated the result is zero. This can be seen since $x^0 = 1$ and we can write, for example, constant 4 as $4x^0$, then differentiating with respect to x we get:

$$4(0)x^{-1} = 0$$

If, as an alternative method, we plot the graph of $y = 4$ we get a straight line parallel with the x-axis as shown in Fig. 10.6.

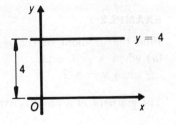

The gradient of the line is zero; that is

$$\dfrac{dy}{dx} = 0.$$

Fig. 10.6

To differentiate an expression containing the sum of several terms, differentiate each individual term separately.

EXAMPLES 6

If $y = 3x^2 + 2x + 3$,

then $\dfrac{dy}{dx} = 3(2)x + 2(1)x^0 + 0 = 6x + 2$

If $\quad\quad y = ax^3 + bx^2 + cx + d$ where a, b, c and d are constants,

then $\quad\dfrac{\mathrm{d}y}{\mathrm{d}x} = 3ax^2 + 2bx + c$

If $\quad\quad y = \sqrt{x} + \dfrac{1}{\sqrt{x}} = x^{1/2} + x^{-1/2}$,

then $\quad\dfrac{\mathrm{d}y}{\mathrm{d}x} = \tfrac{1}{2}x^{-1/2} + (-\tfrac{1}{2})x^{-3/2} = \dfrac{1}{2\sqrt{x}} - \dfrac{1}{2\sqrt{x^3}}$

If $\quad\quad y = 3.1x^{1.4} - \dfrac{3}{x} + 5 = 3.1x^{1.4} - 3x^{-1} + 5$

then $\quad\dfrac{\mathrm{d}y}{\mathrm{d}x} = (3.1)(1.4)x^{0.4} - 3(-1)x^{-2} = 4.34x^{0.4} + \dfrac{3}{x^2}$

So far our differentiation has been in terms of x and y only. But they are only letters representing variables and we may choose other letters or symbols.

Thus, if $\quad s = 3t^2 - 1.5t + 3$,

then $\quad\dfrac{\mathrm{d}s}{\mathrm{d}t} = 6t - 1.5$

The example which follows shows the application of differentiation for finding the gradient of a curve.

EXAMPLE 7

Find the gradient of the graph $y = 3x^2 - 3x + 4$:
(a) when $x = 3$, and,
(b) when $x = -2$.

The gradient at a point is expressed by $\dfrac{\mathrm{d}y}{\mathrm{d}x}$.

If $\quad\quad y = 3x^2 - 3x + 4$,

then $\quad\dfrac{\mathrm{d}y}{\mathrm{d}x} = 6x - 3$

When $\quad x = 3$,

$\quad\quad\dfrac{\mathrm{d}y}{\mathrm{d}x} = 6(3) - 3 = 15$

When $\quad x = -2$,

$\quad\quad\dfrac{\mathrm{d}y}{\mathrm{d}x} = 6(-2) - 3 = -15$

Exercise 18

Differentiate the folllowing:·

1) $y = x^2$

2) $y = x^7$

3) $y = 4x^3$

4) $y = 6x^5$

5) $s = 0.5t^3$

6) $A = \pi R^2$

7) $y = x^{1/2}$

8) $y = 4x^{3/2}$

9) $y = 2\sqrt{x}$

10) $y = 3\sqrt[3]{x^2}$

11) $y = \dfrac{1}{x^2}$

12) $y = \dfrac{1}{x}$

13) $y = \dfrac{3}{5x}$

14) $y = \dfrac{2}{x^3}$

15) $y = \dfrac{1}{\sqrt{x}}$

16) $y = \dfrac{2}{3\sqrt{x}}$

17) $y = \dfrac{5}{x\sqrt{x}}$

18) $s = \dfrac{3\sqrt{t}}{5}$

19) $K = \dfrac{0.01}{h}$

20) $y = \dfrac{5}{x}$

21) $y = 4x^2 - 3x + 2$

22) $s = 3t^3 - 2t^2 + 5t - 3$

23) $q = 2u^2 - u + 7$

24) $y = 5x^4 - 7x^3 + 3x^2 - 2x + 5$

25) $s = 7t^5 - 3t^2 + 7$

26) $y = \dfrac{x + x^3}{\sqrt{x}}$

27) $y = \dfrac{3 + x^2}{x}$

28) $y = \sqrt{x} + \dfrac{1}{\sqrt{x}}$

29) $y = x^3 + \dfrac{3}{\sqrt{x}}$

30) $s = t^{1.3} - \dfrac{1}{4t^{2.3}}$

31) $y = \dfrac{3x^3}{5} - \dfrac{2x^2}{7} - \sqrt{x}$

32) $y = 0.08 + \dfrac{0.01}{x}$

33) $y = 3.1x^{1.5} - 2.4x^{0.6}$

34) $y = \dfrac{x^3}{2} - \dfrac{5}{x} + 3$

35) $s = 10 - 6t + 7t^2 - 2t^3$

36) Find the gradient of the curve $y = 3x^2 + 7x + 3$ at the points where $x = -2$ and $x = 2$.

37) Find the gradient of the curve $y = 2x^3 - 7x^2 + 5x - 3$ at the points where $x = -1.5$, $x = 0$ and $x = 3$.

38) Find the gradient of the graph $2x^2 - 5$ at the points where $x = 1$ and $x = +2$.

39) Find the gradient of the curve $y = x^2 - 3x + 2$ at the point where $x = 3$.

40) For what values of x is the gradient of the curve

$$y = \frac{x^3}{3} + \frac{x^2}{2} - 33x + 7$$

equal to 3?

41) For what values of x is the gradient of the curve $y = 3 + 4x - x^2$ equal to (a) -1,   (b) 0,   (c) 2?

42) If $y = x - \dfrac{1}{x}$ find $\dfrac{dy}{dx}$ where $x = 1$.

TO FIND THE RATE OF CHANGE OF sin θ

The rate of change of a curve at any point is the gradient of the tangent at that point. We shall, therefore, find the gradient at various points on the graph of sin θ and then plot the values of these gradients to obtain a new graph.

It is suggested that the reader follows the method given, plotting his own curves on graph paper.

First, we plot the graph of $y = \sin \theta$ from $\theta = 0°$ to $\theta = 90°$ using values of sin θ obtained from tables which are:

θ	0°	15°	30°	45°	60°	75°	90°
$y = \sin\theta$	0	0.259	0.500	0.707	0.866	0.966	1.000

These values are shown plotted in Fig. 10.7.

Consider point P on the curve, where $\theta = 45°$, and draw the tangent APM. As shown in Example 1 of this chapter, we can find the gradient of the tangent by constructing a suitable right angled triangle AMN (which should be as large as conveniently possible for accuracy) and finding the value of $\dfrac{MN}{AN}$.

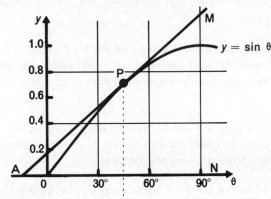

Fig. 10.7

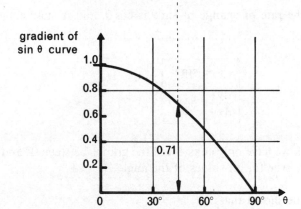

Fig. 10.8

Using the scale on the y-axis MN = 1.29 by measurement and using the scale on the θ-axis AN = 104° by measurement.

In calculations of this type it is necessary to measure AN in radians.

Remember that $\qquad\qquad 360° = 2\pi$ radians

$\therefore \qquad\qquad\qquad\qquad 1° = \dfrac{2\pi}{360}$ radians

$\therefore \qquad\qquad\qquad 104° = \dfrac{2\pi}{360} \times 104 = 1.81$ radians

Hence, $\qquad$ the gradient at P $= \dfrac{MN}{AN} = \dfrac{1.29}{1.81} = 0.71$

The value 0.71 is used as the y-value at $\theta = 45°$ to plot a point on a new graph using the same scales as before. This new graph could be plotted on the same axes as $y = \sin \theta$ but for clarity it has been shown on new axes shown in Fig. 10.7.

This procedure is repeated for points on the $\sin \theta$ curve at θ values 0°, 15°, 30°, 60°, 75° and 90° and the new curve obtained will be as shown in Fig. 10.8. This is the graph of the gradients of the sine curve at various points.

If we now plot a graph of $\cos \theta$, taking values from tables, on the axes in Fig. 10.8 we shall find that the two curves coincide — any difference will be due to errors from drawing the tangents.

Hence the gradient of the $\sin \theta$ curve at any value of θ is the same as the value of $\cos \theta$.

In other words the rate of change of $\sin \theta$ is $\cos \theta$, and in mathematical notation:

If

then

$$y = \sin \theta$$
$$\frac{dy}{dx} = \cos \theta$$

In the above work we have only considered the graphs between 0° and 90° but the results are true for all values of the angle.

Similarly it may be shown that:

If

then

$$y = \cos \theta$$
$$\frac{dy}{dx} = -\sin \theta$$

Exercise 19

Differentiate the expressions in Questions 1–10 each with respect to the appropriate variable.

1) $y = 4 \sin x$ **3)** $y = x - 4 \sin x$

2) $y = 4 + \sin x$ **4)** $y = 3 \sin x - 2$

5) $i = 3 \cos \theta$

6) $v = t + 4 \cos t$

7) $y = \theta^2 - 2 \cos \theta$

8) $i = 2 \cos \theta + 3 \sin \theta$

9) $h = 4t^{2.4} - \dfrac{\cos t}{4} + 16$

10) $p = \dfrac{10}{v} + 6 \sin v$

11)   Show that the following expressions never decrease:

(a) $y = 2x + \cos x$ (b) $i = \sin t + t$

12)   Find the gradient of the curve $y = \cos \theta + \sin \theta$ at each of the following points.

(a) $\theta = 0$ (b) $\theta = \dfrac{\pi}{4}$ (c) $\theta = \pi$

FUNCTIONS

When two variable quantities, x and y, are so related that each value of x is related to exactly one value of y, thus the value of y depends on the value of x, then y is said to be a *function* of x.

We can write this as $y = f(x)$.

When $y = 3x^2 + 4x - 7$ and $x = 5$ then $y = 3(5^2) + 4(5) - 7 = 88$.

Since the value of y depends on the value allocated to x, y is a function of x.

Also $\dfrac{dy}{dx} = 6x + 4$. The expression $6x + 4$ is another function of x known as the *first derived function* or the *first derivative*. We can write
$$f(x) = 3x^2 + 4x - 7$$
$$\Rightarrow f'(x) = 6x + 4$$

If

then

$$\boxed{\begin{array}{c} y = f(x) \\[2mm] \dfrac{dy}{dx} = f'(x) \end{array}}$$

Both methods of indicating a derivative are used.

Thus if $i = 4t^2 + 10$ we say that i is a function of t. That is the value of i depends on the value chosen for t.

Hence $i = f(t)$

and $\dfrac{di}{dt} = 8t$ or $f'(i) = 8t$

FUNCTION OF A FUNCTION

A function like $y = (2x-6)^3$ is a function of a function for $(2x-6)$ is a function of x and $(2x-6)^3$ is a function of $(2x-6)$. Another example of a function of a function is $i = 20 \sin(20t-0.3)$ where i is a function of the sine of an angle which itself is a function of time.

To differentiate a function of a function it is often helpful if we substitute a single variable for the 'inner' function.

DIFFERENTIATION OF A FUNCTION OF A FUNCTION BY SUBSTITUTION

The letter U is often substituted for the 'inner' function when differentiating a function of a function together with the formula:

$$\frac{dy}{dx} = \frac{dy}{dU} \times \frac{dU}{dx}$$

The proof of this important formula is beyond the scope of this course but the application is straightforward and is best shown by example.

EXAMPLE 8

Find $\dfrac{dy}{dx}$ if $y = (x^2-x)^9$.

We have $\qquad\qquad y = (x^2-x)^9$

then $\qquad\qquad\quad y = U^9 \qquad$ where $\quad U = x^2-x$

$\Rightarrow \qquad\qquad \dfrac{dy}{dU} = 9U^8 \quad$ and $\quad \dfrac{dU}{dx} = 2x-1$

but $\qquad\qquad \dfrac{dy}{dx} = \dfrac{dy}{dU} \times \dfrac{dU}{dx}$

$\Rightarrow \qquad\qquad \dfrac{dy}{dx} = 9U^8 \times (2x-1)$

The differentiation has now been completed and it only remains to replace U in terms of x by using our original substitution $U = x^2-x$.

Hence $\qquad\qquad \dfrac{dy}{dx} = 9(x^2-x)^8(2x-1)$

EXAMPLE 9

Find $\dfrac{\mathrm{d}i}{\mathrm{d}t}$ if $i = 200 \sin(20t-0.3)$.

We have $\qquad i = 200 \sin(20t-0.3)$

hence $\qquad i = 200 \sin U \qquad$ where $\quad U = 20t-0.3$

$\Rightarrow \qquad \dfrac{\mathrm{d}i}{\mathrm{d}U} = 200 \cos U \quad$ and $\quad \dfrac{\mathrm{d}U}{\mathrm{d}t} = 20$

but $\qquad \dfrac{\mathrm{d}i}{\mathrm{d}t} = \dfrac{di}{dU} \times \dfrac{dU}{dt}$

$\Rightarrow \qquad \dfrac{\mathrm{d}i}{\mathrm{d}t} = 200 \cos U \times 20$

Replacing U gives

$$\frac{\mathrm{d}i}{\mathrm{d}t} = 4000 \cos(20t-0.3)$$

DIFFERENTIATION OF A FUNCTION OF A FUNCTION BY RECOGNITION

Consider $y = (\ \ )^n$ where any function of x can be written inside the bracket.

Differentiating with respect to x gives

$$\frac{\mathrm{d}y}{\mathrm{d}x} = \frac{\mathrm{d}y}{\mathrm{d}(\ \ )} \times \frac{\mathrm{d}(\ \ )}{\mathrm{d}x}$$

Thus to differentiate an expression of the type $(\ \ )^n$ we first differentiate the bracket, treating it as a term similar to x^n. Secondly differentiate the function of x inside the bracket. Finally, to obtain an expression for $\dfrac{\mathrm{d}y}{\mathrm{d}x}$, multiply the two results together.

EXAMPLE 10

Determine $\dfrac{\mathrm{d}y}{\mathrm{d}x}$ if $y = (x^2-5x+3)^5$.

Differentiating the bracket as a whole we have:

$$\frac{\mathrm{d}y}{\mathrm{d}(\ \ )} = 5(x^2-5x+3)^4 \qquad\qquad (1)$$

Differentiating the function inside the bracket, x^2-5x+3, gives

$$\frac{d(\ )}{dx} = 2x-5 \qquad (2)$$

Multiplying the results (1) and (2) together gives:

$$\frac{dy}{dx} = 5(x^2-5x+3)^4(2x-5)$$

$$= 5(2x-5)(x^2-5x+3)^4$$

Hence by recognising the method we can differentiate directly. Consider for example

$$y = (x^2-3x)^7$$

Hence

$$\frac{dy}{dx} = 7(x^2-3x)^6(2x-3)$$

$$= 7(2x-3)(x^2-3x)^6$$

EXAMPLE 11

Find $\dfrac{d\theta}{dt}$ if $\theta = 5\cos\left(2t^2+\dfrac{3\pi}{2}\right)$

In this example the function of a function is composed of a trigonometric function, $5\cos(\ )$, and an algebraic function, $\left(2t^2+\dfrac{3\pi}{2}\right)$.

We differentiate the two functions in this order and then multiply the two results together to get the required expression for $\dfrac{d\theta}{dt}$.

Thus

$$\frac{d\theta}{dt} = -5\sin\left(2t^2+\frac{3\pi}{2}\right)\times 4t$$

$$= -20t\sin\left(2t^2+\frac{3\pi}{2}\right)$$

EXAMPLE 12

Determine $f'(\theta)$ if $f(\theta) = \sin 7\theta$.

$\{f'(\theta) = $ first derived trigonometric function $\times$ first derived algebraic function.$\}$

Hence

$$f'(\theta) = (\cos 7\theta)\times 7$$

$$= 7\cos 7\theta$$

Exercise 20

In each of the following questions find the differential coefficient with respect to the appropriate variable.

1) $y = (x+1)^5$

2) $y = (3x+2)^3$

3) $y = (6+x)^4$

4) $y = (4x-1)^7$

5) $y = (3-x)^3$

6) $y = (2-3x)^3$

7) $y = (2-x)^4$

8) $y = (4-2x)^5$

9) $y = (x^2-1)^4$

10) $y = (1+x^3)^3$

11) $y = (x^4-2)^5$

12) $y = (1-x^5)^2$

13) $i = (2t+3)^{1/2}$

14) $v = (3\theta-1)^{1/3}$

15) $A = (r-1)^{-4}$

16) $\theta = (2-3t)^{-2}$

17) $r = \sqrt{(2h-5)}$

18) $h = \sqrt{(1-4r)}$

19) $y = \dfrac{6}{1-x}$

20) $y = \dfrac{4}{(2x+7)^2}$

21) $y = \sin 4x$

22) $y = \sin(4x+3)$

23) $i = 5 \sin(2\pi t-0.04)$

24) $y = \cos 2x$

25) $y = \cos(2x+3)$

26) $v = 40 \cos(100\pi t)$

27) $i = 5 \cos(100\pi t-0.001)$

28) $i = 5 \cos(3t-0.1)$

29) $v = 250 \sin(2\pi t)$

30) $v = 50 \cos(0.05-2\pi t)$

AREAS

After reaching the end of this chapter you should be able to:-

1. Determine the area of irregular shapes by means of the mid-ordinate rule.
2. Calibrate a planimeter using plain paper.
3. Use a planimeter to determine an irregular area.
4. Determine the average value of a sine waveform for (a) a half-wave (b) a full-wave.
5. Determine the average value of other common waveforms e.g. triangular and square.

UNITS OF LENGTH

The standard measurement of length is the metre (abbreviation: m). This is split up into smaller units as follows:

$$1 \text{ metre (m)} = 10 \text{ decimetres (dm)}$$
$$= 100 \text{ centimetres (cm)}$$
$$= 1000 \text{ millimetres (mm)}$$

For large measurements the kilometre is used and:

$$1 \text{ kilometre (km)} = 1000 \text{ metres (m)}$$

UNITS OF AREA

The area of a figure is measured by finding how many square units it contains. 1 square metre is the area inside a square which has a side of 1 metre (Fig. 11.1).

Similarly 1 square centimetre is the area inside a square whose side is 1 cm, and 1 square millimetre is the area inside a square whose side is 1 mm

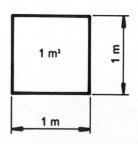

The standard abbreviations for units of area are:

square metres $= \text{m}^2$

square centimetres $= \text{cm}^2$

square millimetres $= \text{mm}^2$

Fig. 11.1

138

Conversions of square units:

$$1 \text{ m}^2 = (100 \text{ cm})^2 = (100 \times 100) \text{ cm}^2 = 10^4 \text{ cm}^2$$

$$1 \text{ m}^2 = (1000 \text{ mm})^2 = (1000 \times 1000) \text{ mm}^2 = 10^6 \text{ mm}^2$$

$$1 \text{ cm}^2 = (10 \text{ mm})^2 = (10 \times 10) \text{ mm}^2 = 10^2 \text{ mm}^2$$

EXAMPLE 1

A room in a factory is 11 m long and 7 m wide. It is to be covered with tiles, the area of each tile being 14 000 mm². How many tiles are needed?

$$\text{Area of room} = 11 \times 7 = 77 \text{ m}^2$$

$$= 77 \times 10^6 \text{ mm}^2$$

and

$$\text{area of tile} = 14\,000 = 14 \times 10^3 \text{ mm}^2$$

$$\therefore \quad \text{number of tiles required} = \frac{\text{area of room}}{\text{area of each tile}}$$

$$= \frac{77 \times 10^6}{14 \times 10^3} = 5.5 \times 10^3 = 5\,500$$

Exercise 21

1)   A sheet of steel 2 m long by 1 m wide is cut up into 400 strips, each strip being 250 m long. What is the width of each strip?

2)   A workshop floor measures 8 m by 6 m. A roll of vinyl floor covering 1500 mm wide and 30 m long is laid. How many square metres of floor are left uncovered?

3)   Blanks of area 20 cm² are being pressed out on a flypress from a sheet of aluminium 1000 mm long and 500 mm wide. Find the area of unused aluminium if 180 blanks are obtained from the sheet.

4)   How many 30 cm square insulation tiles are needed to cover a ceiling 12 m long by 4.5 m wide, allowing for a 5% wastage?

NUMERICAL METHODS FOR CALCULATING IRREGULAR AREAS AND VOLUMES

AREAS

An irregular area is one whose boundary does not follow a definite pattern, e.g. the cross-section of a river.

In these cases practical measurements are made and the results plotted to give a graphical display.

Various numerical methods may then be used to find the area.

USE OF GRAPH PAPER AND 'COUNTING THE SQUARES'

This is probably the simplest method and is often overlooked. It gives results as accurately as those obtained by other more complicated methods.

EXAMPLE 2

A series of soundings taken across the section of a river channel are shown in Fig. 11.2, all dimensions being metres. Find the area of the river cross-section.

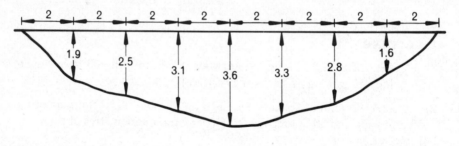

Fig. 11.2

The scales are often chosen to enable the diagram to fit conveniently on a particular size of paper available — the larger the diagram the better.

Fig. 11.3 shows the given results and the plotted points have been joined with a reasonably smooth curve (this is preferable to joining the points with straight lines).

The number of *whole* large squares are then counted — these are shown shaded. The remainder of the area is found by counting the smaller squares, judgement being made for portions of these smaller squares.

Results   Number of large squares = 12

Number of small squares = 173

But on the paper used there are 25 small squares per each large square.

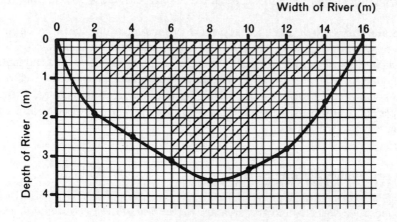

Fig. 11.3

$$\text{Hence the total number of large squares} = 12 + \frac{173}{25}$$

$$= 18.9$$

The scales of the axes must be taken into account:

Consider one large square as shown in Fig. 11.4.

The area of one large square

$$= 2 \text{ m (horizontal)} \times 1 \text{ m (vertical)}$$

$$= 2 \text{ m}^2$$

Represents 2 m

Represents 1 m

Area represents 2 m²

Hence the required area of the river cross-section

$$= 18.9 \times 2 \text{ m}^2$$

$$= 37.8 \text{ m}^2$$

Fig. 11.4

Rough check of result: As in all engineering calculations a rough check should always be made to ensure that the answer obtained is reasonable. This helps to avoid making a big mistake — for instance in this example we may have forgotten to multiply by the scale factor of 2 and obtained an answer one half of the correct value.

The cross-sectional area is approximately equal to that of the rectangle shown in Fig. 11.5,

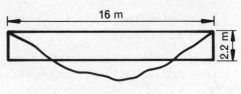

i.e. $16 \text{ m} \times 2.2 \text{ m} = 35.2 \text{ m}^2$

Fig. 11.5

This confirms that the answer obtained is of a reasonable magnitude.

Accuracy of result: The accuracy of the original measurements, the choice of profile when joining the plotted points, and the counting of the incomplete small squares will all affect the final result. This is a difficult example in which to try and obtain a mathematically calculated error. This will be done in examples which follow later in this chapter. From experience, however, we would expect the result obtained to be within an accuracy of $\pm 5\%$.

MID-ORDINATE RULE

Suppose we wish to find the area shown in Fig. 11.6 Let us divide the area into a number of vertical strips, each of equal width b.

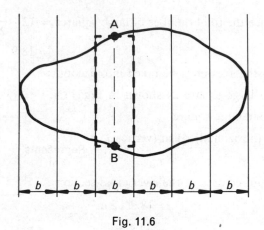

Fig. 11.6

Consider the 3rd strip, whose centre-line is shown cutting the curved boundaries of the area at A and B respectively. Through A and B horizontal lines are drawn and these help to make the dotted rectangle shown. The rectangle has approximately the same area as that of the original 3rd strip, and this area will be $b \times AB$.

AB is called the mid-ordinate of the 3rd strip, as it is mid-way between the vertical sides of the strip.

To find the *whole area*, the areas of the other strips are found in a similar manner and then all are added together for the final result.

$\therefore$ Area = width of strip $\times$ sum of the mid-ordinates.

A useful practical tip to avoid measuring each separate mid-ordinate is to use a strip of paper (Fig. 11.7) and mark off along its edge successive mid-ordinate lengths, as shown in Fig. 11.8. The total area will then be found by measuring the whole length marked out (in the case shown this is HR) and multiplying by the strip width b.

Fig. 11.7

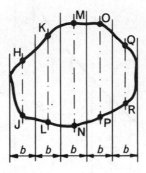

Fig. 11.8

EXAMPLE 3

Find the area under the curve $y = x^2+2$ between $x = 1$ and $x = 4$.

The curve is sketched in Fig. 11.9. Taking 6 strips we may calculate the mid-ordinates.

x	1.25	1.75	2.25	2.75	3.25	3.75
y	3.56	5.06	7.06	9.56	12.56	16.06

Since the width of the strips $= \frac{1}{2}$, the mid-ordinate rule gives:

$$\text{Area} = \tfrac{1}{2} \times (3.56+5.06+7.06+9.56+12.56+16.06).$$

$$= \tfrac{1}{2} \times 53.86 = 26.93 \text{ square units}$$

It so happens that in this example it is possible to calculate an exact answer. How this is done need not concern us at this stage, but by comparing the exact answer with that obtained by the mid-ordinate rule we can see the size of the error.

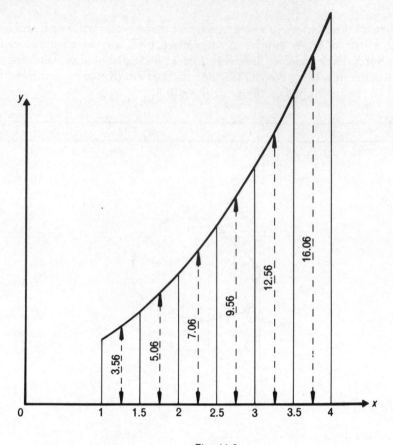

Fig. 11.9

Exact answer = 27 square units

Approximate answer (using the mid-ordinate rule)

= 26.93 square units

∴   Error = 0.07 square units

∴   Percentage error = $\dfrac{0.07}{27} \times 100 = 0.26\%$

From the above it is clear that the mid-ordinate rule gives a good approximation to the correct answer.

THE PLANIMETER

Areas may be measured by means of an instrument called a *planimeter* (Fig. 11.10). It consists of two bars A and B which are hinged at C. Bar B can rotate about a needle point pushed into the paper at D. D is loaded with a weight W to prevent the needle jumping out. Arm A rests on a wheel E, which may roll on the paper as A is moved, and also on an adjustable foot F next to the tracing needle at T.

The tracing point T is taken round the perimeter of the area to be measured and the area read off on the scale attached to wheel E.

The scale has 100 divisions together with an adjacent vernier which enables the scale to be read to one tenth of a division. A small indicator wheel G registers the number of complete revolutions of wheel E (just as the hour hand of a clock records the number of complete revolutions which the minute hand makes).

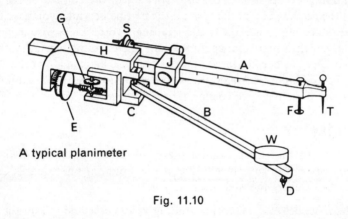

A typical planimeter

Fig. 11.10

The wheel assembly is carried by a portion H whose position is fixed by sliding along bar A before measuring commences. This position is chosen by deciding the units of the required area, e.g. square centimetres, etc. and setting according to the instructions inside the lid of the planimeter case. The instrument should be used on a sheet of drawing paper preferably in a horizontal plane which is sufficiently large to enable the whole of the movement of the wheel E to be made without coming off the paper. The surface of the paper should not be highly polished, otherwise skidding of the wheel may occur (other than that intended) whilst the needle point is being taken round the area perimeter, and hence an incorrect measurement may result.

The tracing point should always be taken clockwise round the boundary — any point may be chosen for the start but the tracing point must always be brought back to that point to complete a measurement. It is best to arrange the initial position so that the arms A and B are approximately at right angles.

This all sounds very complicated but a few minutes using the instrument will show you that it is really quite easy to use, and although you may feel that your hand is not causing the needle to follow the boundary line exactly the errors on each side will usually cancel each other out and a surprisingly accurate measurement can be made after some practice.

An example on the use of the planimeter

Draw a square of side 5 cm. Next set the sliding portion H to the position for measuring square centimetres (see inside the lid of the case). Clamp the small carrier J to the arm A with the pointer on H in approximately the correct position and then set it accurately using the fine adjusting screw S. Adjust the screw foot F so that it takes the weight of the arm of the tracing needle and prevents it catching and sticking in the surface of the paper. Locate the needle point at the spot on the perimeter which you have chosen for the start and take the scale reading. Trace clockwise round the boundary of the square and again take the scale reading when the whole of the boundary has been covered and the needle is back again at the starting point. The difference between these two readings gives the required area, and should of course give 25 cm^2. Check by a second circuit.

Exercise 22

It is suggested that these examples are solved using at least two of the methods covered in the preceding text (i.e. counting the squares, using a planimeter or using the mid-ordinate rule):

1)   The table below gives corresponding values of x and y. Plot the graph and find the area under the graph.

x	1.5	1.7	1.9	2.1	2.3	2.5	2.7	2.9	3.1
y	800	730	622	528	438	366	306	262	214

2)   The table below gives corresponding values of two quantities A and x. Draw the graph and hence find the area under it. (Plot x horizontally.)

A	53.2	35	22.2	21.8	24.2	23.6	18.7	0
x	0	1	2	3	4	5	6	7

3)   Plot the curve given by the following values of x and y and hence find the area included by the curve and the axes of x and y.

x	1	2	3	4	5
y	1	0.25	0.11	0.063	0.040

4)   Plot the curve of $y = 2x^2 + 7$ between $x = 2$ and $x = 5$ and find the area under this curve.

5) Plot the graph of $y = 2x^3 - 5$ between $x = 0$ and $x = 3$ and find the area under the curve.

6) A series of soundings taken across a section of a river channel are given in Fig. 11.11. Find an approximate value for the cross-sectional area of the river at this section.

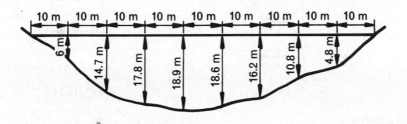

Fig. 11.11

MEAN VALUE

The mean (or average) value or height of a curve is often of importance.

$$\text{The mean value} = \frac{\text{area under the curve}}{\text{length of the base}}$$

A type of graph which is met frequently in electrical engineering is a waveform.

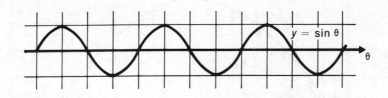

Fig. 11.12

A waveform is a graph which repeats indefinitely. A sine curve (Fig. 11.12) is an example of a waveform.

A portion of the graph which shows the complete shape of the waveform without any repetition is called a cycle.

In the case of the curve of sin θ the portion of the graph over one cycle is said to be a full-wave (Fig. 11.13), and over half of one cycle is said to be a half-wave (Fig. 11.14).

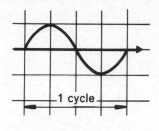

Fig. 11.13   Fig. 11.14

EXAMPLE 4

Find the mean value of sin θ for: (a) a half-wave, (b) a full-wave.

(a)   A half-wave of $y = \sin \theta$ occurs over a range from $\theta = 0°$ to $\theta = 180°$. Values of θ have been taken at 15° intervals and the table of values is:

θ	0°	15°	30°	45°	60°	75°	90°
$\sin \theta$	0	0.259	0.500	0.707	0.866	0.966	1.00

The graph is symmetrical about the vertical line through 90° so the portion from 90° to 180° may be drawn using corresponding values to those for 0° to 90°. The complete curve is shown in Fig. 11.15.

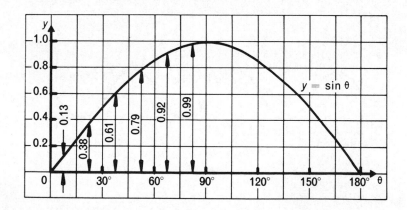

Fig. 11.15

We shall use the mid-ordinate rule to find the area under the curve taking vertical strips of width 15°. The lengths of the mid-ordinates may be found by measurement or alternatively from the sine tables for angles $7\frac{1}{2}°$, $22\frac{1}{2}°$, etc.

Because the curve is symmetrical, as explained above, the area under the curve from 0° to 180° will be twice that from 0° to 90°.

Now, the mean value $= \dfrac{\text{area under the graph}}{\text{length of the base}}$

$$= \frac{2 \times 15 \times (0.13 + 0.38 + 0.61 + 0.79 + 0.92 + 0.99)}{180}$$

$$= 0.64$$

It so happens that in this example it is possible to calculate a more accurate answer. How this is done need not concern us at this stage. The more accurate result is 0.636 6 and this shows that the result we obtained, by using measurements to two decimal places only, is as accurate as could be expected.

(b) A full-wave of $y = \sin \theta$ occurs over a range from $\theta = 0°$ to $\theta = 360°$ The graph will be as shown in Fig. 11.16.

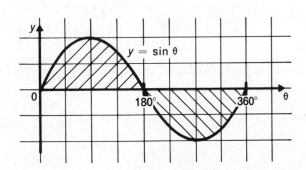

Fig. 11.16

Since the curve is the same shape above and below the horizontal axis the two shaded areas will be equal.

When calculating the mean value the area below the axis must be considered to be negative. This means that the net area under the curve from 0° to 360° will be zero.

Hence, the mean value $= \dfrac{\text{area under the curve}}{\text{length of the base}} = \dfrac{0}{360} = 0$

$\therefore$ the mean value of $\sin \theta$ for a full-wave is zero.

EXAMPLE 5

Calculate the mean value of the square waveform shown in Fig. 11.17.

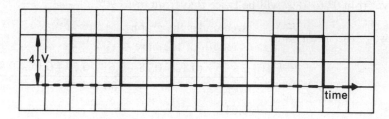

Fig. 11.17

Waveforms are often given on a time base, that is the scale on the horizontal axis is in units of time. However no specific units are given in this problem so we will let t be the time for one cycle as shown in Fig. 11.18.

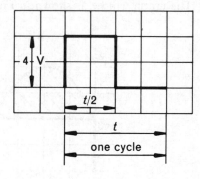

Now, mean $= \dfrac{\text{area under the curve}}{\text{length of the base}}$

$= \dfrac{4(t/2)}{t}$

$= 2\text{ V}$

Hence the mean value is 2 V.

Fig. 11.18

EXAMPLE 6

A ramp wave-form consists of a series of right angled triangles as shown in Fig. 11.19. Find the average height of the wave-form.

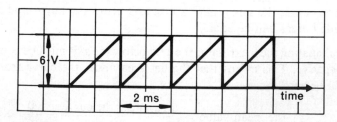

Fig. 11.19

From the diagram the time for one cycle is 2 ms.

but the mean or average height $= \dfrac{\text{area under the graph}}{\text{length of the base}}$

$= \dfrac{\text{area of one triangle}}{\text{base length}}$

$= \dfrac{\frac{1}{2} \times 2 \times 6}{2}$

$= 3 \text{ V}$

ROOT MEAN SQUARE (r.m.s.)

In alternating current work the mean value is not of great importance. This is because we are usually interested in the power produced and this depends on the square of the current values.

The root mean square value $= \sqrt{\text{average height of the } y^2 \text{ curve}}$

i.e. r.m.s. $= \sqrt{\dfrac{\text{area under the } y^2 \text{ curve}}{\text{length of the base}}}$

EXAMPLE 7

Find the r.m.s. value of $\sin \theta$ for: (a) a half-wave, (b) a full-wave.

(a) We shall first find the average height of the y^2 curve for $y = \sin \theta$ in a similar manner to Example 4 in this chapter.

We first draw up a table of values (see over):

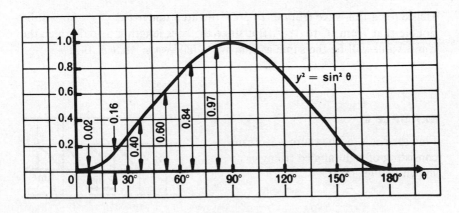

Fig. 11.20

	0°	15°	30°	45°	60°	75°	90°
$y = \sin \theta$	0	0.259	0.500	0.707	0.866	0.966	1.000
$y^2 = \sin^2\theta$	0	0.067	0.250	0.500	0.750	0.933	1.000

The y^2 curve is shown plotted in Fig. 11.20 and the lengths, found by measurement, of the mid-ordinates are as indictaed.

Using the mid-ordinate rule to find the area under the curve we have: The area under the y^2 curve

$$= 2 \times 15 \times (0.02 + 0.16 + 0.40 + 0.60 + 0.84 + 0.97)$$

$$= 2 \times 15 \times 2.99$$

Hence: r.m.s. $= \sqrt{\dfrac{\text{area under the } y^2 \text{ curve}}{\text{length of the base}}}$

$$= \sqrt{\dfrac{2 \times 15 \times 2.99}{180}}$$

$$= 0.71$$

It is possible to calculate a more accurate answer by another method which need not concern us at this stage. The more accurate result is 0.707 1 which confirms our less accurate figure.

(b)   For a full-wave from 0° to 360° the values of $y = \sin \theta$ from 180° to 360° are negative. However when they are squared to give $y^2 = \sin^2\theta$ they will all become positive and give a curve identical to that from 0° to 180°.

Hence for a full-wave from 0° to 360° the area under the y^2 graph will be double that from 0° to 180°. But since the base length is also doubled the r.m.s. value will be the same as that for a half-wave, that is, 0.71.

EXAMPLE 8

Find the r.m.s. of the ramp waveform consisting of right angled triangles as shown in Fig. 11.21.

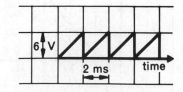

Fig. 11.21

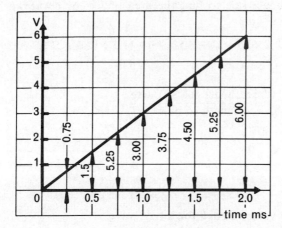

Fig. 11.22

Fig. 11.22 shows one cycle of the ramp wave-form drawn to scale and the voltages, found by measurement, are as shown in the diagram. They are used to draw up a the following table of values:

time ms	0	0.25	0.50	0.75	1.00	1.25	1.50	1.75	2.00
V	0	0.75	1.50	2.25	3.00	3.75	4.50	5.25	6.00
V^2	0	0.56	1.25	5.06	9.00	14.06	20.25	27.56	36.00

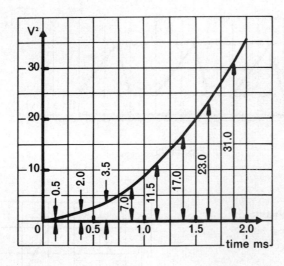

Fig. 11.23

The y^2 curve is shown plotted in Fig. 11.23 and the lengths, found by measurement of the mid-ordinates are as indicated.

Using the mid-ordinate rule to find the area under the curve we have:
the Area under the y^2 curve

$$= 0.25(0.5+2.0+3.5+7.0+11.5+17.0+23.0+31.0)$$

$$= 23.9$$

Hence: r.m.s. $= \sqrt{\dfrac{\text{area under the } y^2 \text{ curve}}{\text{length of base}}}$

$$= \sqrt{\dfrac{23.9}{2}} = 3.46$$

Exercise 23

1) Find the mean value of $\cos\theta$ for a half-wave.

2) Find the average height of the triangular wave-form shown in Fig. 11.24.

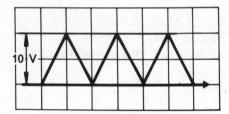

Fig. 11.24

3) Find the mean value of the waveform shown in Fig. 11.25.

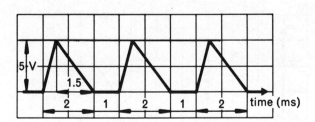

Fig. 11.25

4) Find the mean value of the waveform whose shape for one cycle is as shown in Fig. 11.26.

Fig. 11.26

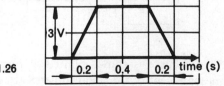

5) Find the r.m.s. value of $\cos \theta$:
(a) for a half-wave,
(b) for a full-wave.

6) Find the r.m.s. value of the waveform shown in Fig. 11.24.

7) Find the r.m.s. value of the waveform shown in Fig. 11.25.

8) Find the r.m.s. value of the waveform shown in Fig. 11.26.

SUMMARY

a) The mid-ordinate rule gives:

$$\text{Area} = (\text{width of strip}) \times (\text{sum of the mid-ordinates})$$

b) The mean value $= \dfrac{\text{area under the curve}}{\text{length of the base}}$

c) The root mean square value $= \sqrt{\dfrac{\text{area under the } y^2 \text{ curve}}{\text{length of the base}}}$

Self-Test 4

1) An area is 5 m². Hence it is:
    a 50 cm²   **b** 500 cm²   **c** 5000 cm²   **d** 50 000 cm²

2) An area is 2000 mm². Hence it is:
    a 2 cm²   **b** 20 cm²   **c** 0.2 cm²   **d** 200 cm²

3) An area is 600 000 mm². Hence it is:
    a 6000 m²   **b** 600 m²   **c** 6 m²   **d** 0.6 m²

4) An area is 0.3 m². Hence it is:
    a 30 mm²   **b** 300 mm²   **c** 30 000 mm²   **d** 300 000 mm²

5) The mean value of the waveform shown in Fig. 11.27 over a full wave is:
    a 2 V   **b** 0.5 V   **c** 1 V   **d** $\dfrac{4}{3}$ V

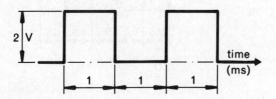

Fig. 11.27

6) The r.m.s. value of the waveform shown in Fig. 11.27 over a full wave is:

    a 2 V   **b** $\dfrac{8}{3}$ V   **c** 1.5 V   **d** $\sqrt{2}$ V

7) The mean value of the waveform shown in Fig. 11.28 over a full wave is:

    a 2 V   **b** 1 V   **c** 0 V   **d** -1 V

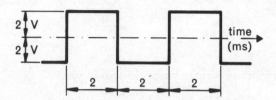

Fig. 11.28

8) The r.m.s. value of the waveform shown in Fig. 11.28 over a full wave is:

    a 2 V   **b** $\sqrt{2}$ V   **c** 1 V   **d** $\dfrac{1}{\sqrt{2}}$ V

9) A planimeter is used for measuring:

    a lengths   **b** perimeters   **c** areas   **d** volumes

10) The area of the diagram shown in Fig. 11.29 using the mid-ordinate rule is:

    a $0.5 \times \frac{1}{2}(1+1.2+1.6+2.2+2.6+2.8)$
    b $0.5 \times (1+1.2+1.6+2.2+2.6+2.8)$
    c $5 \times 0.5 \times (1.1+1.4+1.9+2.4+2.7)$
    d $0.5 \times (1.1+1.4+1.9+2.4+2.7)$

Fig. 11.29

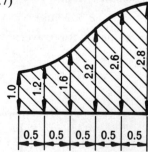

12. BOOLEAN ALGEBRA

INTRODUCTION

Since 1940 Boolean algebra has played an increasingly significant role in the design of switching circuits and other two-state devices. The two states are usually represented by the symbols 0 and 1.

In computers and microprocessors binary signals are used. These signals can only have two levels whose actual values depend upon the circuit components. For instance, a transistor circuit might use the levels -3 volts (to represent 0) and $+3$ volts (to represent 1).

The various circuits in a machine like a computer use logic elements or gates such as NOT, AND and OR gates in various combinations. These gates combine and manipulate the binary signals so that the required functions are performed.

SET THEORY

Since every Boolean algebra may be interpreted as an algebra of sets, for some choice of universal set, the following notation and laws must be known.

1) The elements of a set are denoted by $a,b,c, \ldots$, and sets by $A,B,C, \ldots$

2) When listing the elements of a set we use { } notation.

3) Elements *not* in set A are denoted by $\bar{A}$. Thus NOT $A = \bar{A}$.

4) If the set A is contained in the set B we write $A \subset B$. If at the same time $B \subset A$ then $A = B$.

5) The set containing *all* the elements under discussion is called the *universal set* and is denoted by E (entirety), U (universal) or by the symbol 1. In this book the symbol 1 is used and it should not be confused with the number 1.

6) The set containing *no* elements is called the *null* set and is denoted by the symbols { }, $\emptyset$ or 0. In this book the symbol 0 is used and it should not be confused with the number 0.

7) A set containing 3 elements has 2^3 subsets; that is 8 subsets. A set containing 4 elements has 2^4 subsets and so on.

COMBINATION OF SETS

Union

The *union* of two sets is denoted by $+$ and we define

$$A+B = \{x: x \epsilon A \text{ or } x \epsilon B \text{ or } x \epsilon \text{ both sets.}\}$$

EXAMPLE 1

$A = \{0,1,2,3\}$ and $B = \{3,4,5\}$.

Hence $\qquad\qquad A+B = \{0,1,2,3,4,5\}$

Intersection

The *intersection* of two sets is denoted by $.$ and we define

$$A.B = \{x: x \epsilon A \text{ and } x \epsilon B\}$$

EXAMPLE 2

$A = \{0,1,2,3\}$ and $B = \{3,4,5\}$.

Hence $\qquad\qquad A.B = \{3\}$

LAWS OF SET THEORY APPLIED TO THE OPERATORS + AND .

The following laws are very helpful and the results should be memorised. Each law can be verified using a Venn diagram.

Law	$+$	$\cdot$
1. Tautology	$A+A = A$	$A.A = A$
2. Absorption	$A+A.B = A$	$A.(A+B) = A$
3. Complementation	$A+\bar{A} = 1$	$A.\bar{A} = 0 \quad \bar{\bar{A}} = A$
4. De Morgan	$(\overline{A+B}) = \bar{A}.\bar{B}$	$\overline{A.B} = \bar{A}+\bar{B}$
5. Commutative	$A+B = B+A$	$A.B = B.A$
6. Associative	$A+(B+C) = (A+B)+C$	$A.(B.C) = (A.B).C$
7. Distributive	$A.(B+C) = A.B+A.C$	$A+B.C$ $= (A+B).(A+C)$

Operations with 0 and 1

(a) $1+A = 1$ (b) $0.A = 0$ (c) $0+A = A$

(d) $1.A = A$ (e) $\bar{1} = 0$ (f) $\bar{0} = 1$

Duality

Examination of the above laws shows that in any identity in the algebra of sets if we replace each . by + and each + by ., each 0 by 1 and each 1 by 0, then the resulting expression is also true.

EXAMPLE 3

The distributive law states $A.(B+C) = A.B+A.C$.

Replacing each ., +, 0 and 1 as explained above gives

$$A+(B.C) = (A+B).(A+C)$$

which is the other distributive law.

Simplification of Sets

The methods usually employed are:
1) Expansion and simplification.
2) Factorisation.

EXAMPLE 4

a) Expand and simplify the expression $X+\bar{X}.Y$. Use a Venn diagram to verify the result.

$$X+\bar{X}.Y = (X+\bar{X}).(X+Y) \quad \dots \text{ Distributive law}$$
$$= \quad 1 \quad .(X+Y)$$
$$= X+Y$$

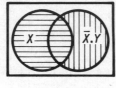 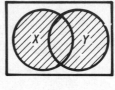

$$\qquad X+\bar{X}.Y \qquad\qquad\qquad\qquad X+Y$$

b) Simplify the expression $A(\bar{A}+B)+B(B+C)+B$.

$$A(\bar{A}+B)+B(B+C)+B = A\bar{A}+AB+BB+BC+B$$
$$= 0+AB+B+BC+B$$
$$= B+AB+BC$$
$$= B(1+A+C)$$
$$= B.1$$
$$= B$$

With practice many of the intermediate steps can be eliminated. Note that AB should be written $A.B$ and terms like $B(B+C)$ should be written $B.(B+C)$. We assume the . to be in position.

c) Simplify $\overline{XY+\bar{X}Y+X\bar{Y}}$ to a single term and verify the result by drawing a Venn diagram.

$$\overline{XY+\bar{X}Y+X\bar{Y}} = (\bar{X}+\bar{Y})(X+\bar{Y})(\bar{X}+Y) \quad \dots \text{ De Morgan's law}$$
$$= (\bar{X}+\bar{Y})(X\bar{X}+XY+\bar{X}\bar{Y}+\bar{Y}Y)$$
$$= (\bar{X}+\bar{Y})(XY+\bar{X}\bar{Y})$$
$$= \bar{X}XY+\bar{X}\bar{Y}+XY\bar{Y}+\bar{X}\bar{Y}$$
$$= \bar{X}\bar{Y}$$

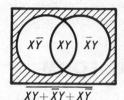

$$\overline{XY+\bar{X}Y+X\bar{Y}}$$

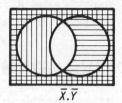

$$\bar{X}.\bar{Y}$$

Using cross-hatching or colour pencils we see that both expressions represent the same area.

Exercise 24

1) Given that $A = \{3,5,7,9,11\}$, $B = \{3,4,5,6,7,8\}$, $C = \{2,4,6,8,10\}$ and that the universal set is $\{2,3,4,5,6,7,8,9,10,11\}$, find:

(a) $A+B$ (b) $B.C$ (c) $\bar{A}.C$ (d) $A.(B+C)$
(e) $\overline{A.(B+C)}$ (f) $\bar{A}+\bar{B}.\bar{C}$ (g) $\overline{A}.\overline{B}$ (h) $\overline{A}.\overline{B}$

2) Use the laws of set theory to simplify each of the following expressions to a single term.

(a) $A.\bar{B}.\bar{A}.\bar{B}$ (b) $AB+A\bar{B}+\bar{A}B+\bar{A}\bar{B}$ (c) $A\bar{C}+ABC+AC$
(d) $(X+Y)(X+Y)$ (e) $(X+XY+XYZ)(X+Y+Z)$

3) Draw Venn diagrams to show the equivalence of the following results.

(a) $\overline{X.Y} = \bar{X}+\bar{Y}$ (b) $\bar{X}.\bar{Y} = \overline{X+Y}$
(c) $(X\bar{Y}+YZ)(\bar{X}\bar{Y}+XZ+YZ) = XZ+YZ$

4) Expand and simplify the following set expressions.

(a) $(A+B)(\bar{A}+B)(A+\bar{B})(\bar{A}+\bar{B})$
(b) $X+YW+(W+Y)X+(Z+W)(X+Y)$

CONSTANTS

A constant is something (a number, a quantity, etc.) which has a fixed value. There are a great many possible constants including numbers such as 7, $\frac{3}{4}$, 0.58 and π. In Boolean algebra, however, there are only two possible constants. These are 0 and 1.

VARIABLES

A variable is a quantity, usually denoted by a letter, which can change by taking any value in the system being discussed. Thus, at any one time a variable may take the value of a constant and at a later time the value of some other constant. Since in Boolean algebra there are only two constants, 0 and 1, a variable can become either 0 or 1.

DISJUNCTIVE NORMAL FORM

A *minterm* is a Boolean variable which is the product of all the variables being considered; each one occurring with or without a bar. For example $AB\bar{C}$, ABC and $\bar{A}BC$ are minterms in the three variables A, B and C. $A\bar{B}$ is *not* a minterm in the three variables A, B and C.

A *maxterm* is a Boolean variable which is the sum of all the variables being considered; each one occurring with or without a bar. For example $(\bar{X}+Y+Z)$ and $(X+\bar{Y}+\bar{Z})$ are both maxterms in the three variables X, Y and Z.

A Boolean function is in *disjunctive normal form* when it is expressed as a *sum of minterms*.

A useful technique which enables us to change a term into a minterm, uses the fact that $(X+\bar{X}) = 1$. For example, it can be used if we are working with the two variables X and Y and have the function

$$f = \bar{X}Y+X \qquad \text{(the term } X \text{ is not a minterm)}$$
$$= \bar{X}Y+X(Y+\bar{Y}) \quad \text{(that is } X.1 = X)$$
$$= \bar{X}Y+XY+X\bar{Y}$$

We have now expressed the function as a sum of minterms.

EXAMPLE 5

Express $f = (\overline{A\bar{B}+A\bar{C}})+\bar{A}$ in disjunctive normal form.

Now
$$f = (\overline{A\bar{B}+A\bar{C}})+\bar{A}$$
$$= (\bar{A}+B)(\bar{A}+C)+\bar{A} \quad \ldots \text{ De Morgan's law}$$
$$= \bar{A}+\bar{A}B+\bar{A}C+BC+\bar{A}$$
$$= \bar{A}+\bar{A}B+\bar{A}C+BC$$
$$= \bar{A}(1+B+C)+BC$$
$$= \bar{A}+BC$$

We have now expressed f as the sum of two terms, neither of them minterms. By multiplying $\bar{A}$ by $(B+\bar{B})$ and $(C+\bar{C})$, and multiplying BC by $(A+\bar{A})$ we introduce the missing variables without altering values.

Hence

$$f = \bar{A} + BC$$
$$= \bar{A}(B+\bar{B})(C+\bar{C}) + BC(A+\bar{A}) \quad \ldots \text{(i.e. } \bar{A}.1.1 + BC.1)$$
$$= \bar{A}BC + \bar{A}B\bar{C} + \bar{A}\bar{B}C + \bar{A}\bar{B}\bar{C} + ABC + \bar{A}BC$$
$$= \bar{A}BC + \bar{A}B\bar{C} + \bar{A}\bar{B}C + \bar{A}\bar{B}\bar{C} + ABC$$

which is of the required form.

TRUTH TABLES

Let us consider the variables A and B. Now both A and B can take the values 0 and 1. There are, therefore, four possible combinations for A and B. These are shown in tabular form below.

A	B
0	0
0	1
1	0
1	1

To make a truth table for the expression $A+B$ we add a third column as shown below.

A	B	$A+B$
0	0	0
0	1	1
1	0	1
1	1	1

By looking at the truth table above we see that when $A = 0$ and $B = 1$, the value of $A+B = 1$.

Working in a similar way we can make a truth table for the expression $A.B$.

A	B	$A.B$
0	0	0
0	1	0
1	0	0
1	1	1

When truth tables for more complicated expressions are required, inter-mediate columns may be placed in the truth table. Thus to make a truth table for the expression $A+\bar{B}$ we introduce an intermediate column for $\bar{B}$ as shown below.

A	B	$\bar{B}$	$A+\bar{B}$
0	0	1	1
0	1	0	0
1	0	1	1
1	1	0	1

EXAMPLE 6

Make a truth table for the expression $\bar{A}.B.\bar{C}$

A	B	C	$\bar{A}$	B	$\bar{C}$	$\bar{A}.B.\bar{C}$
0	0	0	1	0	1	0
0	0	1	1	0	0	0
0	1	0	1	1	1	1
0	1	1	1	1	0	0
1	0	0	0	0	1	0
1	0	1	0	0	0	0
1	1	0	0	1	1	0
1	1	1	0	1	0	0

From this truth table we see that the only time the output is 1 is when $A = 0$, $B = 1$ and $C = 0$.

Exercise 25

1) Write a truth table for the expression $\bar{A}+B$.

2) Write a truth table for the expression $\bar{A}.\bar{B}$.

3) Write a truth table for the expression $A+\bar{A}+B$.

4) Write a truth table for the expression $A+\bar{B}+C$.

5) Write a truth table for the expression $A.\bar{B}.C$.

EQUIVALENT EXPRESSIONS

A truth table may be used to find if two expressions are equivalent.

EXAMPLE 7

Use a truth table to prove that $A+\bar{A}=1$

A	$\bar{A}$	$A+\bar{A}$
0	1	1
1	0	1

We see from the truth table that for both values of A, $A+\bar{A}=1$ and hence we have proved that $A+\bar{A}=1$.

Including the method shown in Example 7, we can now establish Boolean relationships in three ways; using the laws of the algebra of sets, drawing Venn diagrams or compiling a truth table.

EXAMPLE 8

Use a truth table to establish the identity

$$(X+Y)(\bar{X}+Z)(Y+Z) = (X+Y)(\bar{X}+Z)$$

X Y Z	$(X+Y)(\bar{X}+Z)(Y+Z) =$		$(X+Y)(\bar{X}+Z) =$	
0 0 0	0 × 1 × 0	0	0 × 1	0
0 0 1	0 × 1 × 1	0	0 × 1	0
0 1 0	1 × 1 × 1	1	1 × 1	1
0 1 1	1 × 1 × 1	1	1 × 1	1
1 0 0	1 × 0 × 0	0	1 × 0	0
1 0 1	1 × 1 × 1	1	1 × 1	1
1 1 0	1 × 0 × 1	0	1 × 0	0
1 1 1	1 × 1 × 1	1	1 × 1	1

The two results columns in the table are the same; hence the identity is established.

FUNCTIONS FROM A TRUTH TABLE

If we know the values in the results column of a truth table then we can write down the function which produced the values. We can either write down the function as a *sum of minterms* from those values which give 1, or as a *product of maxterms* from those values which give 0.

EXAMPLE 9

Determine, and then simplify, the three variable function which is specified by the following truth table.

A	B	C	Result $=f$
0	0	0	0
0	0	1	1
0	1	0	0
0	1	1	0
1	0	0	0
1	0	1	1
1	1	0	1
1	1	1	0

Reading the rows, in the table, opposite the 1's in the results column we have:

$$f = \bar{A}\bar{B}C + A\bar{B}C + AB\bar{C}$$
$$= \bar{B}C(\bar{A} + A) + AB\bar{C}$$
$$= \bar{B}C + AB\bar{C}$$

Instead of reading the rows opposite the 1's if we take the rows opposite the 0's we obtain a product of maxterms.

Thus $f = (\bar{A} + \bar{B} + \bar{C})(\bar{A} + B + \bar{C})(\bar{A} + B + C)(A + \bar{B} + \bar{C})(A + B + C)$

Although this is an equivalent expression the sum of minterms, in this case, is obviously better.

COMPLEMENT

The *complement* of a function can be obtained in various ways. We can put the function to be complemented inside a bracket, 'bar' the complete bracket, and then apply De Morgan's rule.

EXAMPLE 10

Complement the function $f = A + \bar{B}C$.

Now $f = A + \bar{B}C$

Hence
$$\bar{f} = (\overline{A + \bar{B}C})$$
$$= \bar{A}.(B + \bar{C})$$

An alternative method is to write down the missing minterms.

EXAMPLE 11

Express $f = A + \bar{A}B$ as a sum of minterms and then complement the function.

Now
$$f = A + \bar{A}B$$
$$= A(B + \bar{B}) + \bar{A}B \quad \ldots \text{(i.e. } A.1 + \bar{A}B)$$
$$= AB + A\bar{B} + \bar{A}B$$

The missing minterm is $\bar{A}\bar{B}$.

Hence
$$\bar{f} = \bar{A}\bar{B}$$

A third method of obtaining the complement of a function is to rewrite 1's as 0's and 0's as 1's in the result column of a truth table and then write down the complementary function as a sum of minterms.

Exercise 26

1) Use a truth table to show that the two functions $f_1 = (\overline{\bar{A}B + A\bar{C} + \bar{B}C})$ and $f_2 = ABC + \bar{A}\bar{B}\bar{C}$ have equivalent output expressions.

2) Express each of the following as a sum of minterms.

(a) $A + \bar{A}\bar{B}$ (b) $X\bar{Y} + \bar{X}Z + XYZ$ (c) $XYZ + (X + Y)(X + Z)$

3) Write as a sum of minterms, each of the three functions f_1, f_2 and f_3 as specified in the following truth table. Simplify the results.

X	Y	Z	f_1	f_2	f_3
1	1	1	0	0	1
1	1	0	1	1	0
1	0	1	0	1	0
1	0	0	0	0	1
0	1	1	0	0	1
0	1	0	1	1	0
0	0	1	0	1	0
0	0	0	0	0	1

4) Write down the complements of the functions in Question 2 by considering the missing minterms.

THE SYSTEMATIC SIMPLIFICATION OF BOOLEAN FUNCTIONS

Many methods exist but the most popular method is the Veitch–Karnaugh (V–K) method. In 1952 Veitch developed a method of 'mapping' Boolean functions which in 1953 was modified into a more convenient form by Karnaugh.

Every V–K map contains 2^n squares for n variables; each square corresponding to a particular minterm. Various forms of maps are used but the following is the most usual.

1) Two variables, say A and B.
$n = 2$ hence $2^2 = 4$ squares.

2) Three variables, A, B and C.
$n = 3$ hence $2^3 = 8$ squares.
(Note the change in order of the numbers 11 and 10 in numbering the squares.)

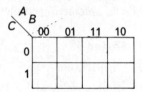

3) Four variables, A, B, C and D.
$n = 4$ hence $2^4 = 16$ squares.

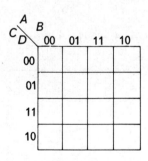

How a Boolean function is plotted on to a V–K map is best illustrated by an example.

EXAMPLE 12

Plot the Boolean function $f = \bar{A}B + BC$ on to a V–K map.

Since there are three variables we require a map with eight squares as shown in 2) above.

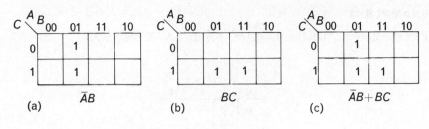

Fig. 12.1

Fig. 12.1(a) shows the map of the term $\bar{A}B$. We place a 1 in the two possible squares which can represent $\bar{A}B$. Two squares are required since $\bar{A}B$ is not a minterm and C is undefined.

Fig. 12.1(b) shows the two possible positions for BC (here A is undefined) and Fig. 12.1(c) shows (a) and (b) combined to give the required map.

EXAMPLE 13

Plot on a V–K map the Boolean expression $XY\bar{Z}+WX+\bar{W}XYZ$.

This is a four-variable function and so $2^4 = 16$ squares are required. The four drawings of Fig. 12.2 show how each part is plotted until the required result is achieved in Fig. 12.2(d).

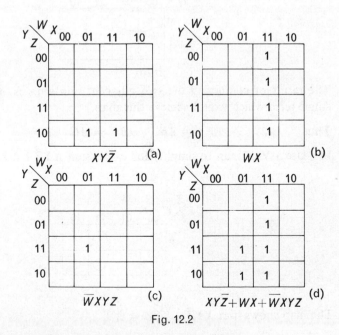

Fig. 12.2

EXAMPLE 14

Plot the Boolean function $f = AB\bar{C} + \bar{A}$ on a V–K diagram.

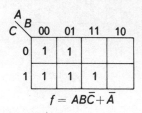

$$f = AB\bar{C} + \bar{A}$$

COMBINING TERMS

To simplify the expressions represented in 2, 3 and 4 variable V–K maps, two minterms may be combined if they are *adjacent* to each other in the same row or column, or if they are at *opposite ends* of the same row or column.

EXAMPLE 15

a) Use a V–K map to simplify the Boolean function $f = \bar{A}\bar{B}\bar{C} + \bar{A}B\bar{C}$.

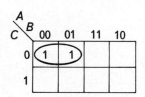

The result of combining the two adjacent minterms is $\bar{A}\bar{C}$. This is the single term which would achieve this map.

Thus $\bar{A}\bar{B}\bar{C} + \bar{A}B\bar{C} = \bar{A}\bar{C}$

b) Use a V–K map to simplify the expression $XY\bar{Z} + XYZ$.

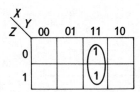

The map shows that $XY\bar{Z} + XYZ = XY$

c) Simplify the Boolean function $f = \bar{A}\bar{B}C + B\bar{C} + A\bar{B}C$ with the aid of a V–K diagram.

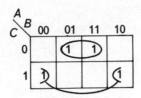

By combining appropriate pairs the map displays an equivalent function

$$f = B\bar{C} + \bar{B}C$$

Some of the other patterns of combinations of minterms are shown in Fig. 12.3 and the following points should be noted:

1) Each 1–square in a V–K diagram must be used *at least* once.

2) Each 1–square may be used as often as desired.

3) Generally 1–square groups should be as large as possible.

4) The number of 1–squares in a group must *always be a power of* 2.

5) Since a choice of grouping is often available there is usually more than one 'best' answer.

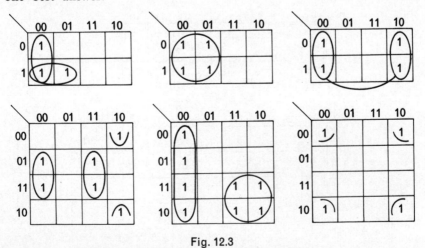

Fig. 12.3

EXAMPLE 16

Use a four-variable V–K map to simplify $f = \bar{B}C + \bar{A}B\bar{C} + ACD + B\bar{C}D + ABD$

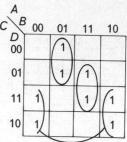

Combining the squares as shown gives the simpler equivalent result:

$$f = \bar{B}C + \bar{A}B\bar{C} + ABD$$

Exercise 27

1)   Construct a two-variable V–K map for each of the following Boolean expressions.

(a) $X + \bar{Y} + X\bar{Y}$ (b) $A + B + \bar{A}\bar{B}$ (c) $\bar{X}Y + X\bar{Y}$

2)   Construct a three-variable V–K map for each of the following Boolean functions. First expand the function when necessary.

(a) $f = X + \bar{Y} + \bar{X}Y + \bar{Z}$ (b) $f = (X + \bar{Y} + \bar{X}Y)\bar{Z}$
(c) $f = (A + B)C + A(BC)$

3)   Write down a two-term expression for each of the following V–K maps.

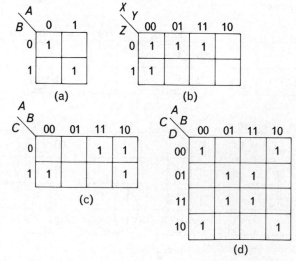

4)   Use an appropriate V–K map to simplify, where possible, the following Boolean functions. First expand the function when necessary.

(a) $f = X + \bar{Y} + \bar{X}YZ$ (b) $f = A + AB + BC$

(c) $f = (X + \bar{Y} + \bar{X}Y)(X + \bar{Y}) + \bar{X}Z$ (d) $f = \bar{A}\bar{B}\bar{C} + A\bar{B}\bar{C} + A\bar{B}C + \bar{A}\bar{B}C$

(e) $f = (XY + \bar{X}Y + X\bar{Y})$ (f) $f = WX + \bar{W}Z + XYZ$

(g) $f = (A + \bar{B} + C)(\bar{A}B + D)$

LOGIC ELEMENTS

An important application of Boolean algebra is in the theory of switching circuits involving two-state (bistable) devices.

We designate each switch by a letter. If two switches are operated so that they open and close simultaneously we designate them by the same letter. If two switches are such that the first is always closed when the second is open, and vice-versa, we designate one by X and the other by $\bar{X}$,(NOT X).

A circuit which consists of two switches, A and B, connected in parallel will be denoted by $A + B$ (A OR B), and a circuit consisting of two switches, A and B, connected in series will be denoted by $A.B$ (A AND B).

Thus to each series–parallel circuit there corresponds an algebraic expression and, conversely, to each algebraic expression, involving only $+$ and $.$, there corresponds a series–parallel circuit.

A closed switch is represented by the symbol 1 and an open switch by the symbol 0. Various types of binary logic elements or gates are used in these transistorised circuits, the most common being the NOT (NEGATER), OR, NOR, AND and NAND gates.

Gate	Symbol	
NOT (NEGATER)	A —[1]o— $\bar{A}$	When the input signal represents 1 the output signal represents 0, and vice versa.
OR	A —[1]— $A+B$ B	The output signal is represented by 1 only when one or more of the input signals is represented by 1.
NOR	A —[1]o— $\overline{A+B}$ B	Output $\overline{A+B} = \bar{A}.\bar{B}$ (De Morgan's law).
AND	A —[&]— $A.B$ B	The output signal is represented by 1 only when ALL input signals are represented by 1.
NAND	A —[&]o— $\overline{A.B}$ B	Output $\overline{A.B} = \bar{A}+\bar{B}$ (De Morgan's law).

EXAMPLE 17

Draw a logic diagram to represent the Boolean function $f = A+BC$.

We require an AND element for the term $B.C$ (B AND C), and an OR element to combine both terms.

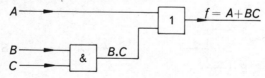

EXAMPLE 18

A particular output can usually be achieved in a variety of ways depending on the availability or cost of individual gates. Use the laws of the algebra of sets to obtain three different outputs each equivalent to the function $f = \bar{A}.\bar{B}+\bar{A}.\bar{C}$ and hence draw the three logic diagrams.

Using De Morgan's law twice we obtain the three different functions each with an equivalent output.

$$f = \bar{A}.\bar{B}+\bar{A}.\bar{C} \qquad \text{(Fig. 12.4(a))}$$

$$f = (\overline{A+B})+(\overline{A+C}) \qquad \text{(Fig. 12.4(b))}$$

$$f = \overline{(A+B)(A+C)} \qquad \text{(Fig. 12.4(c))}$$

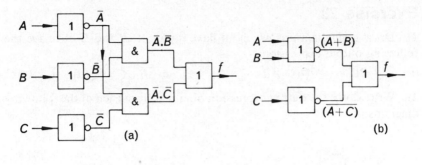

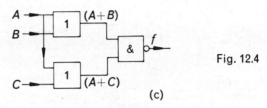

(c)

Fig. 12.4

EXAMPLE 19

The output from a network having three inputs, A, B and C is given by

$$f = \bar{A}\bar{B}\bar{C} + ABC + A\bar{B}\bar{C} + A\bar{B}C$$

Use a Veitch–Karnaugh map to find a simpler equivalent output and represent this output with the three gates, one OR, one NOR and one AND.

From the map, Fig. 12.5(a), the simpler function with an equivalent output,

$$f = AC + \bar{B}\bar{C}$$

is obtained.

In order to represent this output using the specified gates we must replace $\bar{B}.\bar{C}$ with $\overline{B+C}$, using De Morgan's law.

Hence $\qquad\qquad f = AC + \overline{B+C}$ $\qquad$ (See Fig. 12.5(b)).

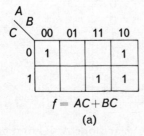

(a)

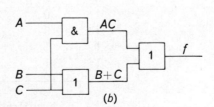

(b)

Fig. 12.5

Exercise 28

1)   Draw logic diagrams which all have inputs A, B and C but have the following output expressions:

(a) $A+BC$ (b) $AB+\bar{C}$ (c) $\overline{A+B}+C$ (d) $\bar{A}+\bar{B}+C$

2)   Write down the output expression obtained from each of the following diagrams:

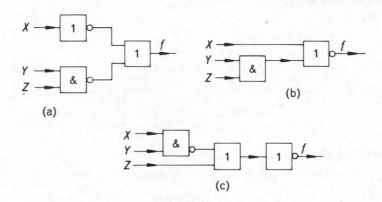

(a)

(b)

(c)

3)   Use a V–K diagram to simplify $f = \bar{A}B\bar{C}+AB\bar{C}+AC$ and then draw the simpler logic circuit with the equivalent output.

4)   Write down, as a sum of minterms, the output expression given by the following truth table. Use a V–K map to show that a simpler but equivalent output is given by the function $f = X\bar{Z}+X\bar{Y}$. Represent this output by a logic diagram using one AND and one NAND. (Remember $\bar{Z}+\bar{Y} = \overline{Z.Y}$.)

X	Y	Z	f
0	0	0	0
0	0	1	0
0	1	0	0
0	1	1	0
1	0	0	1
1	0	1	1
1	1	0	1
1	1	1	0

5)   Draw a logic circuit using one NOT, one OR and one AND to represent the output obtained from the V–K diagram of Fig. 12.6.

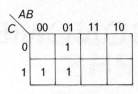

Fig. 12.6

Self-Test 5

1) The NOT gate changes X to:

    a 0   **b** 1   **c** $\bar{X}$ **d** $-X$

2) 0.1.0.1.0.0.1 equals:

    a 1   **b** 0   **c** $\bar{1}$ **d** $\bar{0}$

3) $1.\bar{0}.1.\bar{0}.1$ equals:

    a 1   **b** $\bar{1}$ **c** 0   **d** $\bar{0}$

4) If $\bar{A}.\bar{0}.B = 1$ then A equals:

    a 0   **b** 1   **c** $\bar{0}$ **d** $\bar{1}$

5) $1+\bar{0}+\bar{1}+0+\bar{1}$ is equal to:

    a 1   **b** 0   **c** -1 **d** $\bar{1}$

6) $\bar{M}.\bar{N}$ is equal to:

    a $(M+N)$ **b** $(\overline{M+N})$ **c** $(\bar{M}+\bar{N})$ **d** $(\overline{\bar{M}+\bar{N}})$

7) $\bar{R}+\bar{S}$ is equal to:

    a $\bar{R}.\bar{S}$ **b** $\overline{R}.\overline{S}$ **c** $R.S$ **d** $\overline{R}.\overline{S}$

8) The complement of $AB+\bar{A}\bar{B}+\bar{A}B$ is:

    a $A+B$ **b** $\overline{A+B}$ **c** $\bar{A}+\bar{B}$ **d** $A.\bar{B}$

9) From the truth table of Fig. 12.7 the function f is equal to:

    a $\bar{W}X+W\bar{X}$ **b** $(\bar{W}+\bar{X})(W+X)$ **c** $\bar{W}\bar{X}+WX$ **d** $\bar{W}\bar{X}+\bar{W}X$

10) The Veitch–Karnaugh map of Fig. 12.8 represents:

    a $\bar{A}$ **b** $\bar{A}C$ **c** $\bar{A}\bar{C}$ **d** ABC

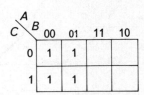

Fig. 12.7   Fig. 12.8

13. TRIGONOMETRICAL GRAPHS

AMPLITUDE OR PEAK VALUE

The graphs of sin θ and cos θ each have a maximum value of $+1$ and a minimum value of -1.

Similarly the graphs of $R.\sin \theta$ and $R.\cos \theta$ each have a maximum value of $+R$ and a minimum value of $-R$. These graphs are shown in Fig. 13.1.

The value of R is known as the amplitude or peak value.

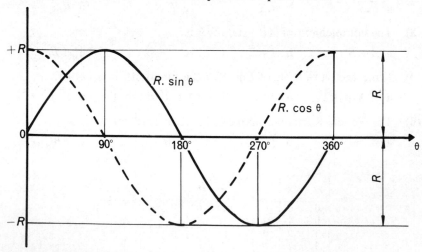

Fig. 13.1

178

EXAMPLE 1

Draw the curve of $E = 100 \sin \theta$.

Using the table of sines we draw up the table shown below.

$\theta°$	0	30	60	90	120	150	180
$\sin \theta$	0	0.500 0	0.866 0	1.000	0.866 0	0.500 0	0
$100 \sin \theta$	0	50	86.6	100	86.6	50	0

$\theta°$	210	240	270	300	330	360
$\sin \theta$	$-0.500\,0$	$-0.866\,0$	-1.000	$-0.866\,0$	$-0.500\,0$	0
$100 \sin \theta$	-50	-86.6	-100	-86.6	-50	0

The curve is shown in Fig. 13.2.

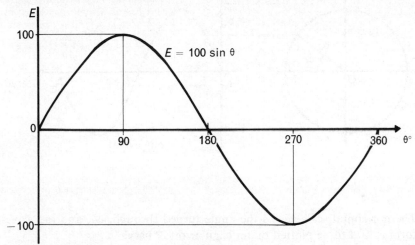

Fig. 13.2

RELATION BETWEEN ANGULAR AND TIME SCALES

In Fig. 13.3 OP represents a radius, of length R, which rotates at a uniform angular velocity ω radians per second about O, the direction of rotation being anticlockwise.

Now $\text{angular velocity} = \dfrac{\text{angle turned through}}{\text{time taken}}$

∴ $\text{angle turned through} = (\text{angular velocity}) \times (\text{time taken})$

and hence after a time t seconds

$\text{angle turned through} = \omega t \text{ radians}$

Also from the right angled triangle OPM:

$$\frac{\text{PM}}{OP} = \sin \hat{POM}$$

∴   PM = OP.sin $\hat{POM}$

or   PM = R.sin ωt

If a graph is drawn, as in Fig. 13.3, showing how PM varies with the angle ωt the sine wave representing R.sin ωt is obtained. It can be seen that the peak value of this sine wave is R (i.e. the magnitude of the rotating radius).

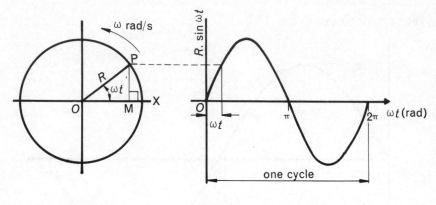

Fig. 13.3

The horizontal scale shows the angle turned through, ωt, and the waveform is said to be plotted on an **angular** or ωt **base**.

CYCLE

A cycle is the portion of the waveform which shows its complete shape without any repetition. It may be seen from Fig. 13.3 that one cycle is completed whilst the radius OP turns through 360° or 2π radians.

PERIOD

This is the time taken for the waveform to complete one cycle.

It will also be the time taken for OP to complete one revolution or 2π radians.

Now we know that time taken = $\dfrac{\text{angle turned through}}{\text{angular velocity}}$

hence

$$\boxed{\text{the period} = \dfrac{2\pi}{\omega} \text{ seconds}}$$

FREQUENCY

The number of cycles per second is called the frequency. The unit of frequency representing one cycle per second is the hertz (Hz).

Now if   1 cycle is completed in $\dfrac{2\pi}{\omega}$ seconds (a period)

then $1 \div \dfrac{2\pi}{\omega}$ cycles are completed in 1 second

and therefore $\dfrac{\omega}{2\pi}$ cycles are completed in 1 second

Hence

$$\boxed{\text{frequency} = \dfrac{\omega}{2\pi} \text{ Hz}}$$

Also since $\text{period} = \dfrac{2\pi}{\omega} \text{ s}$

then

$$\boxed{\text{frequency} = \dfrac{1}{\text{period}}}$$

TIME BASE

We have seen how a graph may be plotted on an 'angular' or 'ωt' base as in Fig. 13.3. Alternatively the units on the horizontal axis may be those of time (usually seconds), and this is called a 'time' base.

RADIANS AND DEGREES

We know that one full revolution is equivalent to $360°$ or 2π radians.

Hence $1 \text{ radian} = \left(\dfrac{360}{2\pi}\right)^{\circ} = \left(\dfrac{180}{\pi}\right)^{\circ}$

If tables are used when finding trigonometrical ratios, such as sines and cosines, of angles it may be necessary to convert an angle in radians to an angle in degrees.

For example $\sin 0.5 = \sin \left(0.5 \times \dfrac{180}{\pi} \right)^{\circ} = \sin 28.65^{\circ} = 0.4795$

It should be noted that if the units of an angle are omitted it is assumed that it is given in radians (as in the above example).

If a scientific calculator is available it is often possible to set the machine to accept radians by setting a special key. There is then no necessity to convert from radians to degrees.

GRAPHS OF $\sin t$, $\sin 2t$, $\sin 3t$, AND $\sin \frac{1}{2}t$

Consider $\sin t$.   Since the period of $\sin \omega t$ is $\dfrac{2\pi}{\omega}$ seconds

then   the period of $\sin t$ is $\dfrac{2\pi}{1} = 6.28$ seconds.

In order to plot one complete cycle of the waveform it is necessary to take values of t from 0 to 6.28 seconds. The reader may find it useful to draw up a suitable table of values and plot the curve. The curve is shown plotted on a time base in Fig. 13.4.

Similarly, the waveform **sin 2t** has a period of $\dfrac{2\pi}{2} = 3.14$ seconds

and   the waveform **sin 3t** has a period of $\dfrac{2\pi}{3} = 2.09$ seconds

and   the waveform **sin ½t** has a period of $\dfrac{2\pi}{\frac{1}{2}} = 12.56$ seconds

All these curves are shown plotted in Fig. 13.4. This enables a visual comparison to be made and it may be seen, for example, that the curve of $\sin 3t$ has a frequency of three times that of $\sin t$ (since three cycles of $\sin 3t$ are completed during one cycle of $\sin t$).

GRAPHS OF $R.\cos \omega t$

The waveforms represented by $R.\cos \omega t$ are similar to sine waveforms, R

being the peak value and $\dfrac{2\pi}{\omega}$ the period. The reader is left to plot these as instructed in Exercise 29 which follows this text.

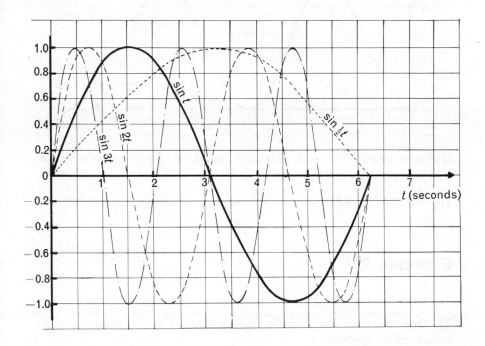

Fig. 13.4

GRAPHS OF sin²t AND cos²t

It is sometimes necessary in engineering applications, such as when finding the root mean square value of alternating currents and voltages, to be familiar with the curves $\sin^2 t$ and $\cos^2 t$.

As previously the period of $\sin t$ is $\dfrac{2\pi}{1} = 6.28$ seconds, and so we will plot the graph of $\sin^2 t$ on a time base using values of t from 0 to 6.28 seconds. Again it may be useful for the reader to construct a suitable table of values and the resulting graph is shown in Fig. 13.5.

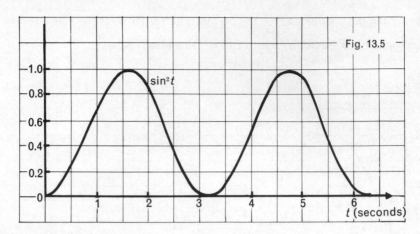

Fig. 13.5

It may be seen that the cycle of the $\sin^2 t$ curve is one half that of $\sin t$, and also the curve is wholly positive.

The reader is left to plot the $\cos^2 t$ curve which is Question 5 of Exercise 29.

Exercise 29

1) Draw the curve of $50.\cos \theta$ over one cycle on an angle base.

2) Draw the curve of $25.\sin 2\theta$ over two cycles on the same axes that were used for Question 1.

3) Find the period of the waveform $\cos t$ and then draw the curve over one cycle on a time base.

4) Draw the waveform represented by $\cos 2t$, $\cos 3t$ and $\cos \frac{1}{2}t$ on the same axes as used for Question 3 and over the period of $\cos t$.

5) Draw the curve of $\cos^2 t$ on a time base over the period of $\cos t$.

EXAMPLE 2

Draw the graph of $5.\sin (2t-1.2)$ over one cycle on a time base.

Now the time for one cycle, that is the period, for $\sin \omega t$ is $\dfrac{2\pi}{\omega}$

$\therefore$ the time for one cycle, that is the period,

for $5.\sin (2t-1.2)$ is $\dfrac{2\pi}{2} = 3.14$ seconds.

It should be noted that the period of sin $(2t-1.2)$ is the same as the period of sin $2t$. As you will see later the figure '-1.2' does not affect the period or frequency of the curve.

Hence the table of values will be from $t = 0$ to $t = 3.14$ seconds and we have chosen 0.2 second intervals.

t	0	0.2	0.4	0.6	0.8	1.0
$2t-1.2$	-1.2	-0.8	-0.4	0	0.4	0.8
sin $(2t-1.2)$	-0.932	-0.717	-0.389	0	0.389	0.717
5.sin $(2t-1.2)$	-4.66	-3.59	-1.95	0	1.95	3.59

t	1.2	1.4	1.6	1.8	2.0
$2t-1.2$	1.2	1.6	2.0	2.4	2.8
sin $(2t-1.2)$	0.932	0.999	0.909	0.675	0.335
5.sin $(2t-1.2)$	4.66	4.99	4.55	3.38	1.67

t	2.2	2.4	2.6	2.8	3.0
$2t-1.2$	3.2	3.6	4.0	4.4	4.8
sin $(2t-1.2)$	-0.058	-0.443	-0.757	-0.952	-0.996
5.sin $(2t-1.2)$	-0.292	-2.21	-3.78	-4.76	-4.98

In addition to the above values it is worth remembering that the curve cuts the axis when angle $(2t-1.2) = 0$ and when $(2t-1.2) = 2\pi$

i.e. when $t = 0.6$ and when $t = 3.74$

It also cuts the axis half way between these two values

i.e. when $t = \dfrac{0.6+3.74}{2} = 2.17$

The maximum peak value $+5$ will occur when the value of t is half way between $t = 0.6$, and $t = 2.17$

i.e. when $t = \dfrac{0.6+2.17}{2} = 1.38$

The minimum peak value -5 will occur when the value of t is half way between $t = 2.17$ and $t = 3.74$

i.e. when $t = \dfrac{2.17+3.74}{2} = 2.95$

The curve is shown plotted in Fig. 13.6.

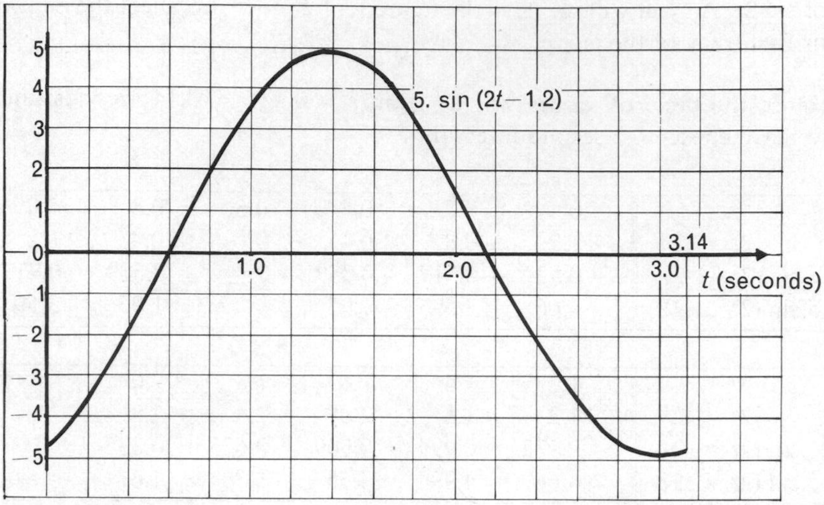

Fig. 13.6

Exercise 30

1) Draw the graph of $\sin(t-0.4)$ over one cycle on a time base.

2) Draw the graph of $\sin(t+0.7)$ over one cycle on a time base.

3) Find the period of the waveform $3.\sin(4t-2)$ and draw the graph over one cycle on a time base.

4) Plot the curve $4.\sin(3t+1.8)$ over one cycle, and find the values of the peak value and the period. Use a time base.

5) Find the equation of the waveform which has a frequency which is twice that of the curve $\sin t$ and has an amplitude of 3. Draw these two curves on the same axes on a time base over the period of $\sin t$.

PHASE ANGLE

The principal use of sine and cosine waveforms occurs in electrical engineering where they represent alternating currents and voltages. In a diagram such as shown in Fig. 13.7 the rotating radii OP and OQ are called phasors.

Fig. 13.7 shows two phasors OP and OQ, separated by an angle α, rotating at the same angular speed in an anti-clockwise direction. The sine waves produced by OP and OQ are identical curves but they are displaced from each other. The amount of displacement is known as the phase difference and, measured along the horizontal axis, is α. The angle α is called the **phase angle**.

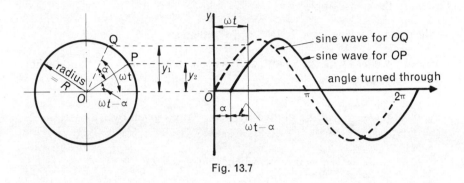

Fig. 13.7

In Fig. 13.7 the phasor OP is said to *lag* behind phasor OQ by the angle α. If the radius of the phasor circle is R then $OP = OQ = R$ and hence:

for the phasor OQ, $\qquad y_1 = R.\sin \omega t$

and for the phasor OP, $\quad y_2 = R.\sin (\omega t - \alpha)$

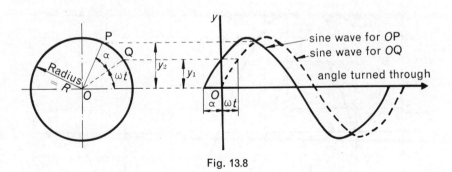

Fig. 13.8

Similarly in Fig. 13.8 the phasor OP leads the phasor OQ by the phase angle, α.

Hence for the phasor OQ, $\quad y_1 = R.\sin \omega t$

and for the phasor OP, $\qquad y_2 = R.\sin (\omega t + \alpha)$

In practice it is usual to draw waveform on an 'angular' or 'ωt' bases, when considering phase angles as in the following example.

EXAMPLE 3

Plot the waveform of $\sin \omega t$ and $\sin\left(\omega t - \dfrac{\pi}{3}\right)$ on an angular base and identify the phase angle.

The cycle of a sine wave is 360° or 2π radians.

Hence $\sin \omega t$ will be plotted between when $\omega t = 0$ and when $\omega t = 2\pi$ radians.

Also $\sin\left(\omega t - \dfrac{\pi}{3}\right)$ will be plotted between values given by:

$$\omega t - \frac{\pi}{3} = 0 \quad \text{and when} \quad \omega t - \frac{\pi}{3} = 2\pi$$

i.e. $$\omega t = \frac{\pi}{3} \quad \text{radians and} \quad \omega t = 2\pi + \frac{\pi}{3} = \frac{7\pi}{3} \text{ radians}$$

The table of values have been omitted, but the reader may find it useful to check these values, and hence the graphs which are shown plotted in Fig. 13.9.

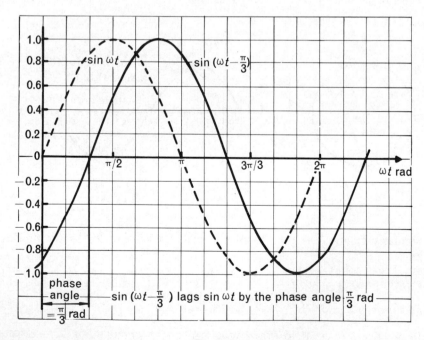

Fig. 13.9

Exercise 31

1) Plot the graphs of sin θ and sin $(\theta+0.9)$ on the same axes on an angular base using units in radians. Indicate the phase angle between the waveforms and explain whether it is an angle of lead or lag.

2) Plot the graph of sin $\left(\omega t+\dfrac{\pi}{6}\right)$ on an angular base over one cycle.

3) Plot the graphs of sin $\left(\omega t+\dfrac{\pi}{3}\right)$ and sin $\left(\omega t-\dfrac{\pi}{4}\right)$ on the same axes on an angular base showing a cycle of each waveform. Identify the phase angle between the curves.

4) Write down the equation of the waveform which:

(a) leads sin ωt by $\dfrac{\pi}{2}$ radians.

(b) lags sin ωt by π radians.

(c) leads sin $\left(\omega t-\dfrac{\pi}{3}\right)$ by $\dfrac{\pi}{3}$ radians.

(d) lags sin $\left(\omega t+\dfrac{\pi}{6}\right)$ by $\dfrac{\pi}{3}$ radians.

COMBINATION OF TWO SINE WAVES

It is sometimes necessary to find the single sine wave resulting from the combination of two sine waves.

Consider the two sine waves:

$$e_1 = A \sin \theta \text{ and } e_2 = B \sin \theta$$

There is no phase difference between the two sine waves and hence the resulting sine wave is:

$$e = (A+B) \sin \theta$$

EXAMPLE 4

Two sinusoidal currents are given by $i_1 = 10 \sin \theta$ and $i_2 = 15 \sin \theta$. Draw the sine wave representing the resultant of these two currents.

The resulting sine wave is given by:

$$i_R = (10+15) \sin \theta = 25 \sin \theta$$

Using a table of sines of angles draw up the table shown below:

$\theta°$	0	30	60	90	120	150	180
$\sin \theta$	0	0.5000	0.8660	1.000	0.8660	0.5000	0
$25 \sin \theta$	0	12.5	21.65	25	21.65	12.5	0

$\theta°$	210	240	270	300	330	360
$\sin \theta$	−0.5000	−0.8660	−1.000	−0.8660	−0.5000	0
$25 \sin \theta$	−12.5	−21.65	−25	−21.65	−12.5	0

The resultant sine wave is shown in Fig. 13.10.

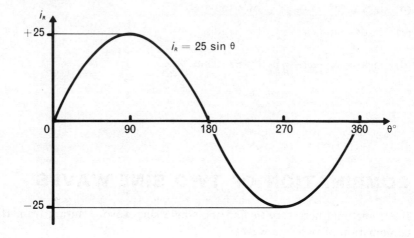

Fig. 13.10

If there is a phase difference between two sine waves then the resulting sine wave may be found by the method shown in Example 5.

EXAMPLE 5

The instantaneous values of two alternating currents are given by: $i_1 = 10 \sin \theta$ and $i_2 = 12 \sin (\theta+30°)$. Draw a graph of the sine wave resulting from a combination of these two currents.

To find the resulting current we draw up the following table:

θ	0	30	60	90	120	150	180
$\sin \theta$	0	0.500 0	0.866 0	1	0.866 0	0.500 0	0
$i_1 = 10 \sin \theta$	0	5	8.66	10	8.66	5	0
$\theta + 30°$	30	60	90	120	150	180	210
$\sin (\theta + 30°)$	0.5	0.866	1	0.866	0.5	0	−0.5
$i_2 = 12 \sin (\theta + 30°)$	6	10.39	12	10.39	6	0	−6
$i_R = i_1 + i_2$	6	15.39	20.66	20.39	14.66	5	−6

θ	210	240	270	300	330	360
$\sin \theta$	−0.500 0	−0.866 0	−1	−0.866	−0.5	0
$i_1 = 10 \sin \theta$	−5	−8.66	−10	−8.66	−5	0
$\theta + 30°$	240	270	300	330	360	390
$\sin (\theta + 30°)$	−0.866	−1	−0.866	−0.5	0	0.5
$i_2 = 12 \sin (\theta + 30°)$	−10.39	−12	−10.39	−6	0	6
$i_R = i_1 + i_2$	−15.39	−20.66	−20.39	−14.66	−5	6

The graphs of i_1, i_2 and i_R are shown in Fig. 13.11 where it will be seen that i_R is a sine wave with a peak value of 21.3 and a phase angle of 16°, and has an equation $i_R = 21.3 \sin (\theta + 16°)$.

The peak value of 21.3 and the phase angle of 16° are only approximate as their accuracy depends on reading values from the scales of the graphs. In this case it is possible to calculate these values by a theoretical method and obtain more accurate results. The method is beyond the scope of this book but the more accurate answers are 21.254 and 16° 24' which shows the answers we obtained are as good as could be expected from a graphical method.

It is possible to obtain the i_R curve from the graphs of i_1 and i_2 by graphical addition. This is, in fact, similar to adding the values of i_1 and i_2 in the table of values.

First plot the graphs of i_1 and i_2 and then the points on the i_R curve may be plotted by adding the ordinates (i.e. the vertical lengths) of the corresponding points on the i_1 and i_2 curves.

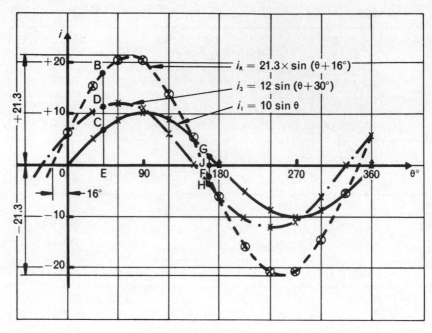

Fig. 13.11

For example to plot point B on the i_R curve in Fig. 13.11 we can measure the lengths CE and DE and then add their values, since BE = CE+DE.

We must take care to allow for the ordinates being positive or negative. For example to find point F on the i_R curve we must use FJ = GJ − HJ.

Exercise 32

1)   Draw the graphs of (a) $y = \sin \theta$ (b) $y = \cos \theta$ for values of θ between 0° and 360°. From the graphs find values of the sin and cos of the angles.

(a) 38°   (b) 72°   (c) 142°   (d) 108°
(e) 200°   (f) 250°   (g) 305°   (h) 328°

2)   Plot the graph of $y = 3 \sin x$ between 0° and 360°. From the graph read off the values of x for which $y = 1.50$ and find the value of y when $x = 250°$.

3)   Draw the graphs of $3 \cos \theta$ for values of θ from 0° to 360°. Use the graph to find approximate values of the two angles for which $3 \cos \theta = 0.6$.

4) By projection from the circumference of a suitably marked off circle draw the graph of $4 \sin \theta$ for values of θ from $0°$ to $360°$. Use the graph to find approximate values of the two angles for which $4 \sin \theta = 1.6$.

5) Plot the voltages given by $v_1 = 2 \sin \theta$ and $v_2 = 4 \sin \theta$ and by graphical addition plot the resultant waveform and find its equation.

6) Find the resultant voltage v_R of the two voltages represented by the equations $v_1 = 3 \sin \theta$ and $v_2 = 5 \sin (\theta - 30°)$ by plotting the three graphs.

7) Plot the graphs of $i_1 = 5 \sin \theta$ and $i_2 = 2 \sin (\theta + 45°)$ and hence find the resultant of i_1 and i_2 by graphical addition. State the equation of the resultant current i_R.

8) The voltages v_1 and v_2 are represented by the equations $v_1 = 30 \sin (\theta + 60°)$ and $v_2 = 50 \sin (\theta - 45°)$. Plot the curves of v_1 and v_2 and the resultant voltage v_R and find the equation representing v_R.

9) Two alternating currents each have instantaneous values given by $i_1 = 16 \sin (5\pi t + 0.7)$ and $i_2 = 28 \sin (5\pi t - 0.4)$ respectively. Compile a table of values of both i_1 and i_2 for values of t between 0 and 0.45 in steps of 0.05. On the same axes draw the graphs of i_1 and i_2. By graphical addition, or otherwise, draw the graph of i_R where $i_R = i_1 + i_2$ and then write down the single expression representing i_R. From the graph of i_R determine the current when $t = 0.022$.

10) A potential difference of $v = 20 \sin \left(100\pi t + \dfrac{\pi}{3} \right)$ occurs when an alternating current of $i = 10 \sin \left(100\pi t - \dfrac{\pi}{6} \right)$ flows through an appliance. Compile a table of values of v and i for values of t between 0 and $\dfrac{1}{60}$, in steps of $\dfrac{1}{600}$. Draw the graphs of v and i. From the graphs determine the values of v and i when $t = 0.004$, and hence calculate the power at this time.

RECIPROCAL RATIOS

In addition to sin, cos and tan there are three other ratios that may be obtained from a right angled triangle. These are:

> cosecant   (called cosec for short)
>
> secant   (called sec for short)
>
> cotangent (called cot for short)

Using the right angled triangle in Fig. 13.12:

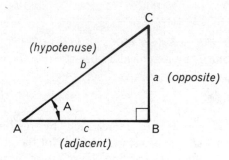

Fig. 13.12

$$\cosec A = \frac{\text{hypotenuse}}{\text{opposite}} = \frac{b}{a}$$

but it is already known that $\sin A = \dfrac{\text{opposite}}{\text{hypotenuse}} = \dfrac{a}{b}$

$\therefore$ $\cosec A = \dfrac{1}{\sin A}$

Similarly,

$$\sec A = \frac{\text{hypotenuse}}{\text{adjacent}} = \frac{b}{c}$$

but it is already known that $\cos A = \dfrac{\text{adjacent}}{\text{hypotenuse}} = \dfrac{c}{b}$

$\therefore$ $\sec A = \dfrac{1}{\cos A}$

and also, $\cot A = \dfrac{\text{adjacent}}{\text{opposite}} = \dfrac{c}{a}$

but it is already known that $\quad \tan A = \dfrac{\text{opposite}}{\text{adjacent}} = \dfrac{a}{c}$

$$\therefore \qquad\qquad \cot A = \dfrac{1}{\tan A}$$

The reciprocal of x is $\dfrac{1}{x}$ and it may therefore be seen why the terms cosec, sec and cot are called 'reciprocal ratios', since they are equal respectively to $\dfrac{1}{\sin}$, $\dfrac{1}{\cos}$ and $\dfrac{1}{\tan}$.

It often helps to simplify calculations if the unknown length in a triangle trigonometry problem is made the numerator of a ratio, and the use of cosec, sec and cot in addition to sin, cos and tan makes this possible.

Tables of values for cosec, sec and cot of angles from 0° to 90° are usually included in books of standard mathematical tables.

These tables are used in a similar way as tables for sin, cos and tan; care must be taken when using the mean differences as to whether the tables give instructions to 'subtract' instead of adding as would normally be done. The examples which follow will explain this procedure.

EXAMPLE 6

Find the length of side x in Fig. 13.13.

Now $\qquad\qquad \dfrac{x}{35} = \operatorname{cosec} 35° 7'$

$\therefore \qquad\qquad x = 35 \times \operatorname{cosec} 35° 7'$

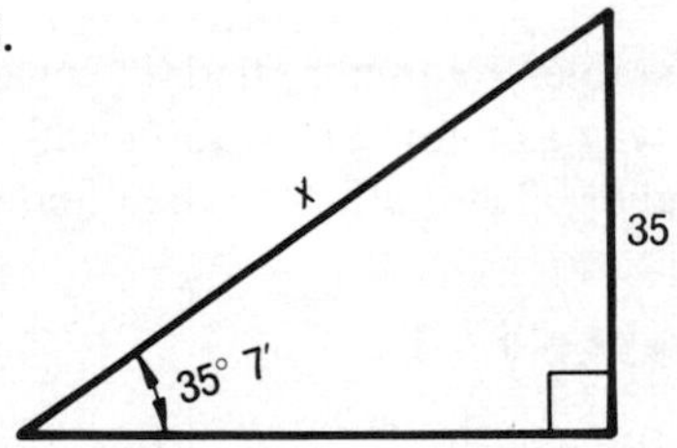

Fig. 13.13

From the table of cosecants, cosec 35° 6′ is read directly as 1.739 1. Looking in the mean difference column headed 1′ the value 7 (representing 0.000 7) is found. The instruction at the head of the table of cosecants indicates that the mean differences must be subtracted:

$\therefore \qquad\qquad \operatorname{cosec} 35° 7' = 1.739\ 1 - 0.000\ 7 = 1.738\ 4$

and hence from above,

$$x = 35 \times 1.738\ 4$$

$$= 60.844 \text{ mm}$$

EXAMPLE 7

Find the length of side y in Fig. 13.14.

$$\text{Now}\ \frac{y}{30} = \sec 64°\ 26'$$

$$\therefore\qquad y = 30 \times \sec 64°\ 26'$$

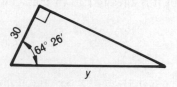

Fig. 13.14

From the table of secants, sec 64° 24′ is read directly as 2.314 4. Looking in the mean difference column headed 2′ the value 28 (representing 0.002 8) is found. Because there is no note to the contrary this value will be added to 2.314 4:

$$\therefore\qquad\qquad \sec 64°\ 26' = 2.314\ 4 + 0.002\ 8 = 2.317\ 2$$

and hence from above,

$$y = 30 \times 2.317\ 2$$

$$= 69.516\ \text{mm}$$

LOGARITHMS OF RECIPROCAL RATIOS

These tables are used to find the logarithm of a trigonometrical ratio in the same way as finding the ratio itself. They save the necessity of looking up the value of the ratio and then finding the corresponding logarithm.

EXAMPLE 8

Find the length of side b in Fig. 13.15.

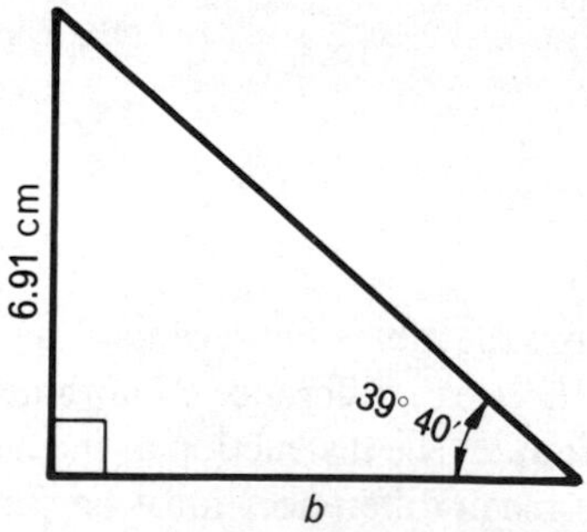

$$\text{Now}\ \frac{b}{6.91} = \cot 39°\ 40'$$

$$\therefore\qquad b = 6.91 \times \cot 39°\ 40'$$

Fig. 13.15

The instruction at the head of the table of log cotangents indicates that the mean differences have to be subtracted.

$\therefore \log \cot 39° 40' = \log \cot (39° 36' + 4')$

$= 0.082\ 3 - 0.001\ 0$

$= 0.081\ 3$

Number	log
6.91	0.839 5
cot 39° 40'	0.081 3
8.333	0.920 8

Hence, using the logarithm calculations shown in the table,

$$b = 8.333 \text{ cm.}$$

Exercise 33

Simple trigonometrical calculations involving the use of cosecant, secant and cotangent.

1) From the tables find the following:

(a) cosec 39° 27'	(b) cosec 67° 23'	(c) sec 11° 7'
(d) sec 49° 28'	(e) cot 37° 49'	(f) cot 74° 11'
(g) log cosec 71° 10'	(h) log cosec 8° 9'	(i) log sec 11° 24'
(j) log sec 29° 3'	(k) log cot 40° 7'	(l) log cot 18° 29'

2) From the tables find the angle θ if:

(a) cosec θ is 1.352 7   (b) sec θ is 1.852   (c) cot θ is 0.491 7

3) Find the lengths of the sides marked x in Fig. 13.16. All the triangles are right angled.

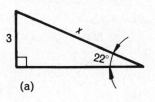

(a)

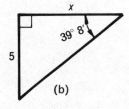

(b)

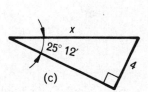

(c)

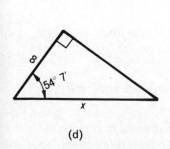

(d)

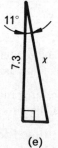

(e)

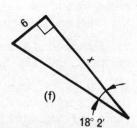

(f)

Fig. 13.16

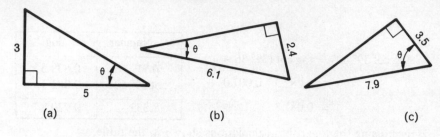

Fig. 13.17

4) By using the cosec, sec or cot find the angles marked θ in the triangles shown in Fig. 13.17. All the triangles are right angled.

5) Calculate the side of the triangle which is marked x in Fig. 13.18.

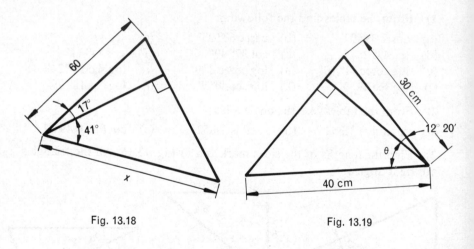

Fig. 13.18   Fig. 13.19

6) Calculate the angle θ in the triangle in Fig. 13.19.

7) The height of an isosceles triangle is 4.3 cm and each of the equal angles is 39°. Find the lengths of the equal sides.

8) Draw a triangle with sides 6 cm, 8 cm, and 10 cm long. Find the cosec, sec and cot of each of the acute angles. Hence find the angles from the tables and check these against your drawing.

9) The chord of a circle is 4.5 cm long and it subtends an angle of 71° at the centre. Calculate the radius of the circle.

10) 20 holes are equally spaced around the circumference of a circle. If the distance between the centres of two adjacent holes, measured along the chord, is 17 mm what is the diameter of the pitch circle?

TRIGONOMETRICAL IDENTITIES

A statement of the type $\cosec A \equiv \dfrac{1}{\sin A}$ is called an IDENTITY. The sign $\equiv$ means 'is identical to'. Any statement using this sign is true for all values of the variables, i.e. the angle A in the above identity. In practice, however, the $\equiv$ sign is often replaced by the $=$ (equals) sign and the identity would be given as $\cosec A = \dfrac{1}{\sin A}$.

Many trigonometrical identities may be verified by the use of a right angled triangle.

EXAMPLE 9

To show that $\tan A = \dfrac{\sin A}{\cos A}$.

The sides and angles of the triangle may be labelled in any way providing that the 90° angle is NOT called A. We have chosen the usual labelling in Fig. 13.20.

Now $\qquad\qquad \sin A = \dfrac{a}{b}$ $\qquad$ Fig. 13.20

and $\qquad\qquad \cos A = \dfrac{c}{b}$

and $\qquad\qquad \tan A = \dfrac{a}{c}$

Hence from the given identity,

R.H. Side $= \dfrac{\sin A}{\cos A} = \dfrac{a/b}{c/b} = \dfrac{a.b}{b.c} = \dfrac{a}{c} = \tan A = $ L.H. Side

EXAMPLE 10

To show that $\cot A = \dfrac{\cos A}{\sin A}$.

Using Fig. 13.20 again we have:

$$\sin A = \dfrac{a}{b}$$

and $\qquad\qquad \cos A = \dfrac{c}{b}$

and $\qquad\qquad \cot A = \dfrac{c}{a}$

Fig. 13.20 *repeat*

Hence from the given identity,

$$\text{R.H. Side} = \frac{\cos A}{\sin A} = \frac{c/b}{a/b} = \frac{c \cdot b}{b \cdot a} = \frac{c}{a} = \cot A = \text{L.H. Side}$$

EXAMPLE 11

To show that $\sin^2 A + \cos^2 A = 1$.

Fig. 13.20 *repeat*

In Fig. 13.20.

$$\sin A = \frac{a}{b} \qquad \therefore \quad \sin^2 A = \left(\frac{a}{b}\right)^2 = \frac{a^2}{b^2}$$

$$\cos A = \frac{c}{b} \qquad \therefore \quad \cos^2 A = \left(\frac{c}{b}\right)^2 = \frac{c^2}{b^2}$$

$$\therefore \qquad \text{L.H.S.} = \sin^2 A + \cos^2 A = \frac{a^2}{b^2} + \frac{c^2}{b^2} = \frac{a^2 + c^2}{b^2}$$

But by Pythagoras' Theorem,

$$a^2 + c^2 = b^2$$

$$\therefore \qquad \text{L.H.S.} = \frac{b^2}{b^2} = 1 = \text{R.H.S.}$$

$$\therefore \qquad \sin^2 A + \cos^2 A = 1$$

EXAMPLE 12

To show that $\sec^2 A = 1 + \tan^2 A$.

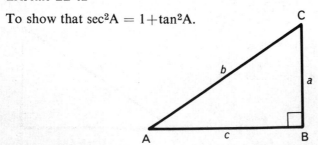

Fig. 13.20 *repeat*

In Fig. 13.20,

$$\sec A = \frac{b}{c} \qquad \therefore \quad \sec^2 A = \left(\frac{b}{c}\right)^2 = \frac{b^2}{c^2}$$

$$\tan A = \frac{a}{c} \qquad \therefore \quad \tan^2 A = \left(\frac{a}{c}\right)^2 = \frac{a^2}{c^2}$$

$$\therefore \qquad R.H.S. = 1 + \tan^2 A = 1 + \frac{a^2}{c^2} = \frac{c^2 + a^2}{c^2}$$

But by Pythagoras' Theorem,

$$a^2 + c^2 = b^2$$

$$\therefore \qquad R.H.S. = \frac{b^2}{c^2} = \sec^2 A = L.H.S.$$

$$\therefore \qquad \sec^2 A = 1 + \tan^2 A$$

EXAMPLE 13

To show that $\cosec^2 A = 1 + \cot^2 A$.

Fig. 13.20   *repeat*

In Fig. 13.20,

$$\cosec A = \frac{b}{a} \qquad \therefore \quad \cosec^2 A = \left(\frac{b}{a}\right)^2 = \frac{b^2}{a^2}$$

$$\cot A = \frac{c}{a} \qquad \therefore \quad \cot^2 A = \left(\frac{c}{a}\right)^2 = \frac{c^2}{a^2}$$

$$\therefore \qquad R.H.S. = 1 + \cot^2 A = 1 + \frac{c^2}{a^2} = \frac{a^2 + c^2}{a^2}$$

But by Pythagoras' Theorem $a^2 + c^2 = b^2$,

$$\therefore \qquad R.H.S. = \frac{b^2}{a^2} = \cosec^2 A = L.H.S.$$

Hence, $\qquad \cosec^2 A = 1 + \cot^2 A$

EXAMPLE 14

If cos A = 0.7 find, without using tables, the values of sin A and tan A, given that $A < \dfrac{\pi}{2}$.

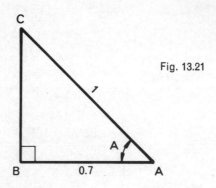

Fig. 13.21

In Fig. 13.21 if we make AB = 0.7 units and AC = 1 unit then

$$\cos A = \frac{0.7}{1} = 0.7.$$

By Pythagoras' Theorem,

$$BC^2 = AC^2 - AB^2 = 1^2 - 0.7^2 = 1 - 0.49 = 0.51$$

$$BC = \sqrt{0.51} = 0.714\,1$$

$$\therefore \qquad \sin A = \frac{BC}{AC} = \frac{0.714\,1}{1} = 0.714\,1$$

and $\qquad \tan A = \dfrac{BC}{AB} = \dfrac{0.714\,1}{0.7} = 1.020$

Exercise 34

Verify the identities in Questions 1–10:

1) $\cos A \sin A = \dfrac{\sin^2 A}{\tan A}$

2) $\sin^2 B \cot B = \sin B \cos B$

3) $\tan^2 C = \dfrac{\sin^2 C}{1 - \sin^2 C}$

4) $\dfrac{1}{\sin^2 D} - 1 = \cot^2 D$

5) $\tan A = \sin A \sec A$

6) $\cos A = \sin A \cot A$

7) $\sec^2 B - 1 = \sin^2 B \sec^2 B$

8) $(\sin A + \cos A)^2 + (\sin A - \cos A)^2 = 2$

9) $\cos^2 C - \sin^2 C = 1 - 2 \sin^2 C$

10) $(1 + \cot^2 A) \sin^2 A = 1$

11)   If $\sin A = 0.317\,3$ and $\cos A = 0.948\,3$ find, without using tables, the value of tan A, given $A < \frac{1}{2}\pi$.

12)   If $\sin A = 0.852\,7$ find the values of cos A and tan A without using tables, given $A < 90°$.

13)   If $\tan A = 1.150\,4$ find the values of sec A, cos A and sin A without using tables, given $A < \frac{1}{2}\pi$.

14)   If $\operatorname{cosec} A = 1.491\,3$, find, without using tables, the values of cot A and also tan A, given $A < \frac{1}{2}\pi$.

15)   Simplify: $\left(1 + \dfrac{1}{\tan^2 A}\right) \sin^2 A.$

AREA OF A TRIANGLE

Three formulae are commonly used for finding the areas of triangles:

1)   If given the base and the altitude (i.e. vertical height).

2)   If given any two sides and the included angle.

3)   If given the three sides.

Case 1)   Given the base and the altitude.

In Fig. 13.22,

Area of triangle $= \frac{1}{2} \times$ **base** $\times$ **altitude**

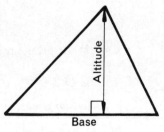

Fig. 13.22

EXAMPLE 15

Find the areas of the triangles shown in Fig. 13.23.

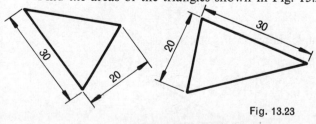

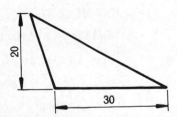

Fig. 13.23

In each case the 'base' is taken as the side of given length and the 'altitude' is measured perpendicular to this side.

Hence   Triangular area $= \frac{1}{2} \times$ base $\times$ altitude

$$= \frac{1}{2} \times 30 \times 20$$

$$= 300 \text{ mm}^2 \quad \text{in each case.}$$

EXAMPLE 16

A trapezium is shown in Fig. 13.24 in which AB is parallel to DC. Find its area.

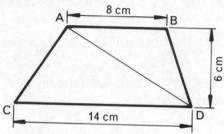

Fig. 13.24

If we join AD then the trapezium is divided into two triangles, the 'bases' and 'altitudes' of which are known.

Hence   area of trapezium $=$ area of $\triangle$ ABD $+$ area of $\triangle$ ADC

$$= \quad \frac{1}{2} \times 8 \times 6 \quad + \quad \frac{1}{2} \times 14 \times 6$$

$$= \quad\quad 24 \quad\quad + \quad\quad 42$$

$$= 66 \text{ cm}^2$$

Case 2)   If given any two sides and the included angle.

In Fig. 13.25,

Area of triangle $= \frac{1}{2}bc \sin A$

or   **area of triangle $= \frac{1}{2}ac \sin B$**

or   **area of triangle $= \frac{1}{2}ab \sin C$**

Fig. 13.25

EXAMPLE 17

Find the area of the triangle shown in Fig. 13.26.

Area $= \frac{1}{2} \times a \times c \times \sin B$

$\quad\quad = \frac{1}{2} \times 4 \times 3 \times \sin 30°$

$\quad\quad = 3 \text{ cm}^2$

Fig. 13.26

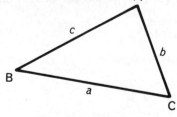

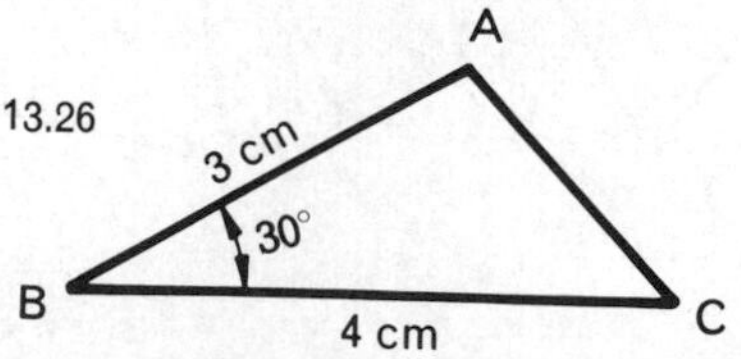

EXAMPLE 18

Find the area of the triangle shown in Fig. 13.27.

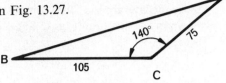

Area $= \frac{1}{2}ab \sin C$

$\qquad = \frac{1}{2} \times 105 \times 75 \times \sin 140°$

Fig. 13.27

We find the value of sin 140° by using the method explained previously as shown in Fig. 13.28, from which it may be seen that

$\qquad \sin 140° = \sin (180° - 140°)$

$\qquad\qquad = \sin 40°$

$\therefore \qquad$ Area $= \frac{1}{2} \times 105 \times 75 \times \sin 40°$

$\qquad\qquad = 2531 \text{ mm}^2$

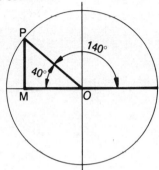

Fig. 13.28

Case 3) If given the three sides.

In Fig. 13.29,

Area of triangle $= \sqrt{s(s-a)(s-b)(s-c)}$

where $\qquad s = \dfrac{a+b+c}{2}$

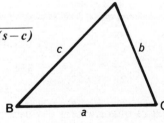

Fig. 13.29

EXAMPLE 19

A triangle has sides of lengths 3 cm, 5 cm and 6 cm. What is its area?

Since we are given the lengths of 3 sides we use

$$\text{area} = \sqrt{s(s-a)(s-b)(s-c)}$$

Now, $\qquad\qquad s = \dfrac{3+5+6}{2} = 7$

$\therefore \qquad\qquad \text{area} = \sqrt{7 \times (7-3) \times (7-5) \times (7-6)}$

$\qquad\qquad\qquad = \sqrt{7 \times 4 \times 2 \times 1}$

$\qquad\qquad\qquad = \sqrt{56} = 7.483 \text{ cm}^2$

EXAMPLE 20

A quadrilateral has the dimensions shown in the diagram (Fig. 13.30). Find its area.

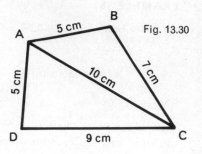

Fig. 13.30

The quadrilateral is made up of the triangles ABC and ACD.

To find the area of $\triangle$ABC.

$$s = \frac{5+7+10}{2} = 11$$

$\therefore$ area of $\triangle$ABC $= \sqrt{s(s-a)(s-b)(s-c)}$

$$= \sqrt{11 \times (11-5)(11-7)(11-10)}$$

$$= \sqrt{11 \times 6 \times 4 \times 1}$$

$$= \sqrt{264} = 16.25 \text{ cm}^2$$

To find the area of $\triangle$ACD,

$$s = \frac{5+9+10}{2} = 12$$

$\therefore$ area of $\triangle$ACD $= \sqrt{s(s-a)(s-b)(s-c)}$

$$= \sqrt{12(12-5)(12-9)(12-10)}$$

$$= \sqrt{12 \times 7 \times 3 \times 2}$$

$$= \sqrt{504} = 22.45 \text{ cm}^2$$

$\therefore$ area of quadrilateral $=$ area of $\triangle$ABC+area of $\triangle$ACD

$$= 16.25 + 22.45 = 38.70 \text{ cm}^2$$

Exercise 35

1) Find the area of a triangle whose base is 7.5 cm and whose altitude is 5.9 cm.

2) Find the area of an isosceles triangle whose equal sides are 8.2 cm and whose base is 9.5 cm.

3) A plate in the shape of an equilateral triangle has a mass of 12.25 kg. If the material has a mass of 3.7 kg/m² find the dimensions of the plate in mm.

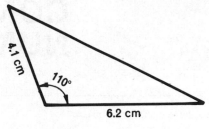

Fig. 13.31

4)   Find the area of the triangle shown in Fig. 13.31.

5)   What is the area of a parallelogram whose base is 7 cm long and whose vertical height is 4 cm?

6)   Obtain the area of a parallelogram if two adjacent sides measure 11.25 cm and 10.5 cm and the angle between them is 49°.

7)   Determine the length of the side of a square whose area is equal to that of a parallelogram with a 3 m base and a vertical height of 1.5 m.

8)   Find the area of a trapezium whose parallel sides are 75 mm and 82 mm long respectively and whose vertical height is 39 mm.

9)   Find the area of a regular hexagon,
(a)   which is 4 cm wide across flats,
(b)   which has sides 5 cm long.

10)   Find the area of a regular octagon,
(a)   which is 2 mm wide across flats,
(b)   which has sides 2 mm long.

11)   The parallel sides of a trapezium are 12 cm and 16 cm long. If its area is 220 cm² what is its altitude?

COMPLEX NUMBERS

On reaching the end of this chapter you should be able to :-

1. *Understand the necessity of extending the number system to include the square roots of negative numbers.*
2. *Define j as $\sqrt{-1}$.*
3. *Define a complex number as consisting of a real part and an imaginary part.*
4. *Define a complex number z in the algebraic form* $x+jy$.
5. *Determine the complex roots of $ax^2+bx+c= 0$ when $b^2 < 4ac$ using the quadratic formula.*
6. *Perform the addition and subtraction of complex numbers in algebraic form.*
7. *Define the conjugate of a complex number in algebraic form.*
8. *Perform the multiplication and division of complex numbers in algebraic form.*
9. *Represent the algebraic form of a complex*

10. *number on an Argand diagram, and show how it may be represented as a phasor.*
11. *Deduce that j may considered to be an operator such that when the phasor representing $x+jy$ is multiplied by j it rotates the phasor through 90° anti-clockwise.*
12. *Understand how phasors on an Argand diagram may be added and subtracted in a manner similar to the addition and subtraction of vectors.*
13. *Show that the full polar form of a complex number is $(\cos \theta+j \sin \theta)$ which may be abbreviated to $r\underline{/\theta}$.*
14. *Perform the operations involved in the conversion of complex numbers in algebraic form to polar form and vice-versa.*
15. *Multiply and divide numbers in polar form.*
16. *Apply the above to problem arising from relevant engineering technology.*

COMPLEX NUMBERS

The solution of the quadratic equation $ax^2+bx+c = 0$ is given by the

formula:
$$x = \frac{-b\pm\sqrt{b^2-4ac}}{2a}$$

When we use this formula most of the quadratic equations we meet, when solving engineering problems, are found to have roots which are ordinary positive or negative numbers.

Consider now the equation $x^2-4x+13 = 0$

then
$$x = \frac{-(-4)\pm(\sqrt{-4)^2-4\times1\times13}}{2\times1}$$

$$= \frac{4\pm\sqrt{16-52}}{2}$$

$$= \frac{4\pm\sqrt{-36}}{2}$$

$$= \frac{4 \pm \sqrt{(-1)(36)}}{2}$$

$$= \frac{4 \pm \sqrt{(-1)} \times \sqrt{(36)}}{2}$$

$$= \frac{4 \pm \sqrt{(-1)} \times 6}{2}$$

$$= 2 \pm \sqrt{-1} \times 3$$

It is not possible to find the value of the square root of a negative number.

In order to try to find a meaning for roots of this type we represent $\sqrt{-1}$ by the symbol j.

(Books on pure mathematics often use the symbol i, but in engineering j is preferred as i is used for the instantaneous value of a current.)

Thus the roots of the above equation become $2+j3$ and $2-j3$.

Definition of a complex number

Expressions such as $2+j3$ are called *complex numbers*. The number 2 is called the *real part* and j3 is called the *imaginary part*.

The general expression for a complex number is $x+jy$, which has a real part equal to x and an imaginary part equal to jy. The form $x+jy$ is said to be *the algebraic form* of a complex number. It may also be called *the cartesian form* or *rectangular notation*.

Powers of j

We have defined j such that

$$j = \sqrt{-1}$$

∴ squaring both sides of the equation

$$j^2 = (\sqrt{-1})^2 = -1$$

Hence
$$j^3 = j^2 \times j = (-1) \times j = -j$$

and
$$j^4 = (j^2)^2 = (-1)^2 = 1$$

and $\qquad\qquad\qquad\qquad j^5 = j^4 \times j = 1 \times j = j$

and $\qquad\qquad\qquad\qquad j^6 = (j^2)^3 = (-1)^3 = -1$

$$\text{and so on.}$$

The most used of the above relationships is $\quad j^2 = -1$.

Addition and subtraction of complex numbers in algebraic form

The real and imaginary parts must be treated separately. The real parts may be added and subtracted and also the imaginary parts may be added and subtracted, both obeying the ordinary laws of algebra.

Thus $\qquad\qquad (3+j2)+(5+j6) = 3+j2+5+j6$

$$= (3+5)+j(2+6) \qquad \cdot$$

$$= 8+j8$$

and $\qquad\qquad (1-j2)-(-4+j) = 1-j2+4-j$

$$= (1+4)-j(2+1)$$

$$= 5-j3$$

EXAMPLE 1

If $\quad z_1$, z_2 and z_3 represent three complex numbers such that $z_1 = 1.6+j2.3$, $\quad z_2 = 4.3-j0.6$ and $\quad z_3 = -1.1-j0.9$ find the complex numbers which represent:

(a) $\quad z_1+z_2+z_3$

(b) $\quad z_1-z_2-z_3$

(a) $\qquad\qquad z_1+z_2+z_3 = (1.6+j2.3)+(4.3-j0.6)+(-1.1-j0.9)$

$$= 1.6+j2.3+4.3-j0.6-1.1-j0.9$$

$$= (1.6+4.3-1.1)+j(2.3-0.6-0.9)$$

$$= 4.8+j0.8$$

(b) $\qquad\qquad z_1-z_2-z_3 = (1.6+j2.3)-(4.3-j0.6)-(-1.1-j0.9)$

$$= 1.6+j2.3-4.3+j0.6+1.1+j0.9$$

$$= (1.6-4.3+1.1)+j(2.3+0.6+0.9)$$

$$= -1.6+j3.8$$

Multiplication of complex numbers in algebraic form

Consider the product of two complex numbers, $(3+j2)(4+j)$.

The brackets are treated in exactly the same way as the rules of algebra

make $(a+b)(c+d) = ac+bc+ad+bd$

Hence $(3+j2)(4+j) = 3\times4+j2\times4+3\times j+j2\times j$

$$= 12+j8+j3+j^2 2$$

$$= 12+j8+j3-2 \qquad \text{since} \quad j^2 = -1$$

$$= (12-2)+j(8+3)$$

$$= 10+j11$$

EXAMPLE 2

Express the product of $2+j$, $-3+j2$, and $1-j$ as a single complex number.

Then $(2+j)(-3+j2)(1-j) = (2+j)(-3+j2+j3-j^2 2)$

$$= (2+j)(-1+j5) \qquad \text{since} \quad j^2 = -1$$

$$= -2-j+j10+j^2 5$$

$$= -7+j9 \qquad \text{since} \quad j^2 = -1$$

Conjugate complex numbers

Consider: $(x+jy)(x-jy) = x^2+jxy-jxy-j^2 y$

$$= x^2-(-1)y^2$$

$$= x^2+y^2$$

Hence we have two product of two complex numbers which produces a real number and therefore does not have a j term. If $x+jy$ represents a complex number then $x-jy$ is known as its *conjugate* (and vice versa). For example the conjugate of $(3+j4)$ is $(3-j4)$ and their product is

$$(3+j4)(3-j4) = 9+j12-j12-j^2 16 = 9-(-1)16$$

$$= 25 \quad \text{which is a real number.}$$

Division of complex numbers in algebraic form

Consider $\dfrac{(4+j5)}{(1-j\ )}$. We use the method of rationalising the denominator.

This means removing the j terms from the bottom line of the fraction. If we multiply $(1-j)$ by its conjugate $(1+j)$ the result will be a real number. Hence, in order not to alter the value of the given expression, we will multiply both the numerator and the denominator by $(1+j)$.

Thus
$$\frac{(4+5j)}{(1-j\ )} = \frac{(4+j5)(1+j)}{(1-j\ )(1+j)}$$

$$= \frac{4+j5+j4+j^2 5}{1-j\ +\ j-j^2}$$

$$= \frac{4+j9+(-1)5}{1-(-1)}$$

$$= \frac{-1+j9}{2}$$

$$= -\frac{1}{2}+j\frac{9}{2}$$

$$= -0.5+j4.5$$

EXAMPLE 3

The impedance Z of a circuit having a resistance and inductive reactance in series is given by the complex number $Z = 5+j6$.

Find the admittance Y of the circuit if $Y = \dfrac{1}{Z}$.

Now
$$Y = \frac{1}{Z} = \frac{1}{5+j6}$$

The conjugate of the denominator is $5-j6$ and therefore multiplying both the numerator and denominator by $5-j6$

then
$$Y = \frac{(5-j6)}{(5+j6)(6-j6)}$$

$$= \frac{5-j6}{25+j30-j30-j^2 36}$$

$$= \frac{5-j6}{25-(-1)36} = \frac{5-j6}{61} = \frac{5}{61}-j\frac{6}{61} = 0.082-j0.098$$

EXAMPLE 4

Two impedances Z_1 and Z_2 are given by the complex numbers $Z_1 = 1+j5$ and $Z_2 = j7$. Find the equivalent impedance Z if:

(a) $Z = Z_1 + Z_2$ when Z_1 and Z_2 are in series.

(b) $\dfrac{1}{Z} = \dfrac{1}{Z_1} + \dfrac{1}{Z_2}$ when Z_1 and Z_2 are in parallel.

(a)
$$Z = Z_1 + Z_2 = (1+j5)+j7$$
$$= 1+j5+j7$$
$$= 1+j12$$

(b)
$$\frac{1}{Z} = \frac{1}{Z_1} + \frac{1}{Z_2} = \frac{1}{(1+j5)} + \frac{1}{j7}$$

$$= \frac{j7+(1+j5)}{(1+j5)j7}$$

$$= \frac{1+j12}{j7+j^2 35}$$

$$= \frac{1+j12}{j7+(-1)35}$$

Thus
$$Z = \frac{j7-35}{1+j12}$$

$$= \frac{(j7-35)(1-j12)}{(1+j12)(1-j12)}$$

$$= \frac{j7-35-j^2 84+j420}{1+j12-j12-j^2 144}$$

$$= \frac{j427-35-(-1)84}{1-(-1)144}$$

$$= \frac{49+j427}{145}$$

$$= 0.338+j2.945$$

Exercise 36

1) Add the following complex numbers:
  (a) $3+j5$, $7+j3$, and $8+j2$
  (b) $2-j7$, $3+j8$, and $-5-j2$
  (c) $4-j2$, $7+j3$, $-5-j6$, and $2-j5$

2) Subtract the following complex numbers:
  (a) $3+j5$ from $2+j8$ (b) $7-j6$ from $3-j9$
  (c) $-3-j5$ from $7-j8$

3) Simplify the following expressions giving the answers in the form $x+jy$:

  (a) $(3+j3)(2+j5)$ (b) $(2-j6)(3-j7)$
  (c) $(4+j5)^2$ (d) $(5+j3)(5-j3)$
  (e) $(-5-j2)(5+j2)$ (f) $(3-j5)(3-j3)(1+j)$
  (g) $\dfrac{1}{2+j5}$ (h) $\dfrac{2+j5}{2-j5}$
  (i) $\dfrac{-2-j3}{5-j2}$ (j) $\dfrac{7+j3}{8-j3}$
  (k) $\dfrac{(1+j2)(2-j)}{(1+j)}$ (l) $\dfrac{4+j2}{(2+j)(1-j3)}$

4) Find the real and imaginary parts of:

  (a) $1+\dfrac{j}{2}$ (b) $j3+\dfrac{2}{j^3}$ (c) $(j2)^2+3(j)^5-j(j)$

5) Solve the following equations giving the answers in the form $x+jy$:

  (a) $x^2+2x+2 = 0$ (b) $x^2+9 = 0$

6) Find the admittance Y of a circuit if $Y=\dfrac{1}{Z}$ where $Z = 1.3+j0.6$.

7) Three impedances Z_1, Z_2, and Z_3 are represented by the complex numbers $Z_1 = 2+j$, $Z_2 = 1+j$, and $Z_3 = j2$. Find the equivalent impedance Z if:

  (a) $Z = Z_1+Z_2+Z_3$ (b) $\dfrac{1}{Z} = \dfrac{1}{Z_1}+\dfrac{1}{Z_2}+\dfrac{1}{Z_3}$

  (c) $Z = \dfrac{1}{\dfrac{1}{Z_1}+\dfrac{1}{Z_2}}+Z_3$

THE ARGAND DIAGRAM

When plotting a graph, cartesian coordinates are generally used to plot the points. Thus the position of the point P (Fig. 14.1) is defined by the coordinates (3, 2) meaning that $x = 3$ and $y = 2$.

Complex numbers may be represented in a similar way on the Argand diagram. The real part of the complex number is plotted along the horizontal real-axis whilst the imaginary part is plotted along the vertical imaginary, or j axis.

However a complex number is denoted, not by a point but, as a phasor, a phasor being a line where regard is paid both to its magnitude and to its direction. Hence in Fig. 14.2 the complex number $4+j3$ is represented by the phasor $\overrightarrow{OQ}$, the end Q of the line being found by plotting 4 units along the real-axis and 3 units along the j-axis.

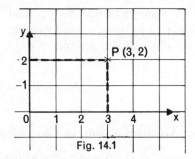
Fig. 14.1

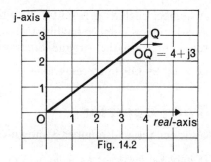
Fig. 14.2

A single letter, the favourite being z, is often used to denote a phasor which represents a complex number. Thus if $z = x+jy$ it is understood that z represents a phasor and not a simple numerical value.

Four typical complex numbers z_1, z_2, z_3, and z_4 are shown on the Argand diagram in Fig. 14.3.

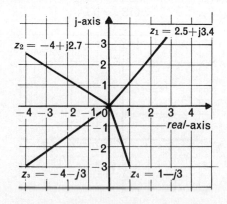
Fig. 14.3

A real number such as 2.7 may be regarded as a complex number with a zero imaginary part, i.e. 2.7+j0, and may be represented on the Argand diagram (Fig. 14.4) as the phasor $z = 2.7$ denoted by $\overrightarrow{OA}$ in the diagram.

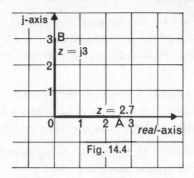

Fig. 14.4

A number such as j3 is said to be wholly imaginary and may be regarded as a complex number having a zero real part, i.e. 0+j3, and may be represented on the Argand diagram (Fig. 14.4) as the phasor $z = j3$ denoted by $\overrightarrow{OB}$ in the diagram.

The j-operator

Consider the real number 3 shown on the Argand diagram, in Fig. 14.5.

It may be regarded as a *phasor*, denoted by $\overrightarrow{OA}$, a phasor being a line where regard is paid both to its magnitude and to its direction. If we now multiply the real number 3 by j we obtain the complex number j3 which may be represented by the phasor $\overrightarrow{OB}$.

It follows that the effect of j on phasor $\overrightarrow{OA}$ is to make it become phasor $\overrightarrow{OB}$, that is $\overrightarrow{OB} = j\overrightarrow{OA}$

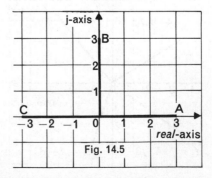

Fig. 14.5

Hence j is known as an operator (called the 'j-operator') which, when applied to a phasor, alters its direction by 90° in an anti-clockwise direction without changing its magnitude.

If we now operate on the vector $\overrightarrow{OB}$ we shall obtain, therefore, vector $\overrightarrow{OC}$.

In equation form this is, $\overrightarrow{OC} = \overrightarrow{jOB}$

but since $\overrightarrow{OB} = \overrightarrow{jOA}$ then $\overrightarrow{OC} = j(\overrightarrow{jOA})$

$$= j^2\overrightarrow{OA}$$

$$= -\overrightarrow{OA} \quad \text{since} \quad j^2 = -1.$$

This is true since it may be seen from the vector diagram that vector $\overrightarrow{OC}$ is equal in magnitude, but opposite in direction, to vector $\overrightarrow{OA}$.

Consider now the effect of the j-operator on the complex number $5+j3$.

In equation form this is: $j(5+j3) = j5+j(j3)$

$$= j5+j^2 3$$

$$= j5+(-1)3$$

$$= -3+j5$$

If phasor $z_1 = 5+j3$ and phasor $z_2 = -3+j5$, it may be seen from the Argand diagram in Fig. 14.6 that their magnitudes are the same but the effect of the operator j on z_1 has been to alter its direction by 90° anti-clockwise to give phasor z_2.

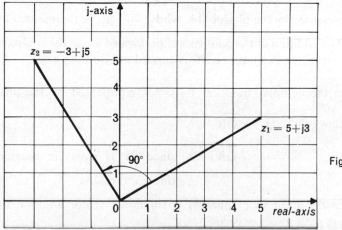

Fig. 14.6

Addition of Phasors

Consider the addition of the two complex numbers   2+j3   and   4+j2.

We have, $(2+j3)+(4+j2) = 2+j3+4+j2$

$$= (2+4)+j(3+2)$$

$$= 6+j5$$

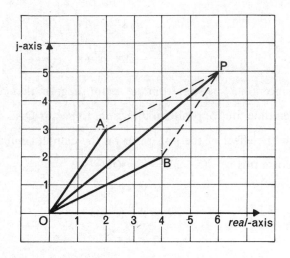

Fig. 14.7

On the Argand diagram shown in Fig. 14.7, the complex number   2+j3   is represented by the phasor $\overrightarrow{OA}$, whilst   4+j2   is represented by phasor $\overrightarrow{OB}$. The addition of the real parts is performed along the *real*-axis and the addition of the imaginary parts is carried out on the j-axis.

Hence the complex number   6+j5   is represented by the phasor $\overrightarrow{OP}$.

It follows that: $\overrightarrow{OP} = \overrightarrow{OB}+\overrightarrow{OA}$

$$= \overrightarrow{OB}+\overrightarrow{BP} \text{ since } \overrightarrow{BP} \text{ is equal in magnitude and direction to } \overrightarrow{OA}.$$

Hence the addition of phasors is similar to vector addition used when dealing with forces or velocities.

Subtraction of Phasors

Consider the difference of the two complex numbers, $4+j5$ and $1+j4$.

We have,

$$(4+j5)-(1+j4) = 4+j5-1-j4$$
$$= (4-1)+j(5-4)$$
$$= 3+j$$

On the Argand diagram shown in Fig. 14.8, the complex number $4+j5$ is represented by the phasor $\overrightarrow{OC}$, whilst $1+j4$ is represented by the phasor $\overrightarrow{OD}$. The subtraction of the real parts is performed along the *real*-axis, and the subtraction of the imaginary parts is carried out along the j-axis. Now let $(4+j5)-(1+j4) = 3+j$ be represented by the phasor $\overrightarrow{OQ}$.

It follows that,

$$\overrightarrow{OQ} = \overrightarrow{OC}-\overrightarrow{OD}$$
$$= \overrightarrow{OC}+\overrightarrow{CQ} \quad \text{since} \quad \overrightarrow{CQ} = -\overrightarrow{OD}.$$

As for phasor addition, the subtraction of phasors is similar to the subtraction of vectors.

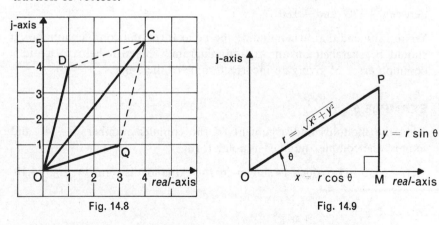

Fig. 14.8   Fig. 14.9

THE POLAR FORM OF A COMPLEX NUMBER

Let z denote the complex number represented by the phasor $\overrightarrow{OP}$ shown in Fig. 14.9. Then from the right angled triangle PMO we have:

$$z = x+jy$$
$$= r\cos\theta+j(r\sin\theta)$$
$$= r(\cos\theta+j\sin\theta)$$

The expression $r(\cos\theta + j\sin\theta)$ is known as *the polar form* of the complex number z. Using conventional notation it may be shown abbreviated as $r\underline{/\theta}$.

r is called the *modulus* of the complex number z and is denoted by   mod z or $|z|$.

Hence, from the diagram, $|z| = r = \sqrt{x^2 + y^2}$
using the theorem of Pythagoras for right-angled triangle PMO.

It should be noted that the plural of *modulus* is *moduli*.

The angle θ is called the *argument* (or amplitude) of the complex number z, and is denoted by   arg z (or amp z).

Hence $\arg z = \theta$

and, from the diagram, $\tan\theta = \dfrac{y}{x}$

There are an infinite number of angles whose tangents are the same, and so it is necessary to define which value of θ to state when solving the equation $\tan\theta = \dfrac{y}{x}$.   It is called the principal value of the angle and lies between $+180°$ and $-180°$.

We recommend that, when finding the polar form of a complex number, it should be sketched on an Argand diagram. This will help to avoid a common error of giving an incorrect value of the angle.

EXAMPLE 5

Find the modulus and argument of the complex number $3+j4$ and express the complex number in polar form.

Let $z = 3+j4$ which is shown in the Argand diagram in Fig. 14.10.

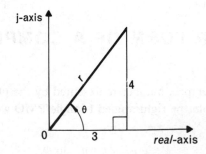

Fig. 14.10

Then $\qquad\qquad |z| = r = \sqrt{3^2+4^2} = 5$

and $\qquad\qquad \tan\theta = \dfrac{4}{3} = 1.3333$

$\therefore \qquad\qquad\qquad \theta = 53°\,8'$

Hence in polar form: $\qquad z = 5(\cos 53°\,8'+\text{j}\sin 53°\,8')$

$$z = 5\underline{/53°\,8'}$$

EXAMPLE 6

Show the complex number $z = 3.5\ \underline{/-150°}$ on an Argand diagram, and find z in algebraic form.

Now z is represented by phasor $\overrightarrow{\text{OP}}$ in Fig. 14.11. It should be noted that since the angle is negative it is measured in a clockwise direction from the *real*-axis datum.

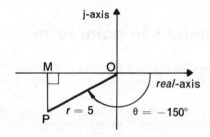

Fig. 14.11 .

In order to express z in algebraic form we need to find the lengths MO and MP. We use the right angled triangle PMO in which
$\text{P}\hat{\text{O}}\text{M} = 180°-150° = 30°$

Now $\qquad\qquad \text{MO} = \text{PO}\cos \text{P}\hat{\text{O}}\text{M} = 5\cos 30° = 4.33$

and $\qquad\qquad \text{MP} = \text{PO}\sin \text{P}\hat{\text{O}}\text{M} = 5\sin 30° = 2.50$

Hence, in algebraic form, the complex number $z = -4.33-\text{j}2.50$.

Multiplying numbers in polar form

Consider the complex number $\quad z_1 = r_1\ \underline{/\theta_1} = r_1(\cos\theta_1+\text{j}\sin\theta_1)$

and another complex number $\quad z_2 = r_2\ \underline{/\theta_1} = r_2(\cos\theta_2+\text{j}\sin\theta_2)$

Then the product of these two complex numbers:

$$z_1 \times z_2 = r_1(\cos\theta_1+\text{j}\sin\theta_1)\times r_2(\cos\theta_2+\text{j}\sin\theta_2)$$

$$= r_1r_2(\cos\theta_1 + j\sin\theta_1)(\cos\theta_2 + j\sin\theta_2)$$

$$= r_1r_2\{\cos\theta_1\cos\theta_2 + j\sin\theta_1\cos\theta_2$$
$$+ j\cos\theta_1\sin\theta_2 + j^2\sin\theta_1\sin\theta_2\}$$

$$= r_1r_2\{(\cos\theta_1\cos\theta_2 - \sin\theta_1\sin\theta_2) +$$
$$j(\sin\theta_1\cos\theta_2 + \cos\theta_1\sin\theta_2)\}$$

$$= r_1r_2\{\cos(\theta_1+\theta_2) + j\sin(\theta_1+\theta_2)\}$$

$$= r_1r_2 \,\underline{/(\theta_1+\theta_2)}$$

Hence to multiply two complex numbers we multiply their moduli and add their arguments.

For example $\qquad 6\,\underline{/17°} \times 3\,\underline{/35°} = 6\times3\,\underline{/17°+35°}$

$$= 18\,\underline{/52°}$$

Dividing numbers in polar form

It can be shown that the division of two complex numbers, using a method similar to that for finding the product of two complex numbers, is given by:

$$\frac{\cdot z_1}{z_2} = \frac{r_1\,\underline{/\theta_1}}{r_2\,\underline{/\theta_2}} = \frac{r_1}{r_2}\,\underline{/\theta_1-\theta_2}$$

Hence we divide two complex numbers we divide their moduli and subtract their arguments.

For example:
$$\frac{5\,\underline{/33°\,55'}}{3\,\underline{/-23°\,40'}} = \frac{5}{3}\,\underline{/(33°\,55')-(-23°\,40')}$$

$$= 1.67\,\underline{/33°\,55'+23°\,40'}$$

$$= 1.67\,\underline{/57°\,35'}$$

EXAMPLE 7

A simple circuit which has a resistance R in series with an inductive reactance X_L has an impedance Z is given by the complex number

$$Z = R + jX_L$$

A simple circuit which has a resistance R in series with a capacitive reactance X_C has an impedance Z given by the complex number

$$Z = R - jX_C$$

Using the above relationships find the resistance and the inductive or capacitive reactance for each of the following impedances:

(a) $8+j12$ (b) $20-j80$ (c) $40\ \underline{/25°}$ (d) $100\ \underline{/-20°}$

(a) Here $Z = 8+j12$, and since it is of the form $Z = R+jX_L$ we can say that the: resistance $R = 8$
and the inductive reactance $X_L = 12$

(b) Here $Z = 20-j80$, and since it is of the form $Z = R-jX_C$ we can say that the: resistance $R = 20$
and the capacitive reactance $X_C = 80$

(c) The complex number $Z = 40\ \underline{/25°}$ is shown on the Argand diagram in Fig. 14.12. If we let $Z = x+jy$, then from the diagram:

$$x = 40\cos 25° \quad \text{and} \quad y = 40\sin 25°$$

$$= 36.3 \qquad\qquad\qquad = 16.9$$

Hence $Z = 36.3+j16.9$ which is of the form $Z = R+jX_L$ and we can say that resistance $R = 36.3$
and the inductive reactance $X_L = 16.9$

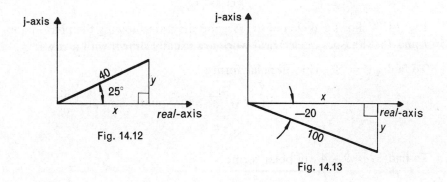

Fig. 14.12

Fig. 14.13

(d) The complex number $Z = 100\ \underline{/-20°}$ is shown on the Argand diagram in Fig. 14.13. If we let $Z = x+jy$, then from the diagram

$$x = 100\cos 20° \quad \text{and} \quad y = 100\sin 20°$$

$$= 94.0 \qquad\qquad\qquad = 34.2$$

but we can see from the diagram that the y value is negative hence $Z = 94.0-j34.2$, which is of the form $Z = R-jX_C$ and we can say that the resistance $R = 94.0$
and the capacitive reactance $X_C = 34.2$.

EXAMPLE 8

The potential difference across a circuit is given by the complex number $V = 50+j30$ volts, and the current is given by the complex number $I = 9+j4$ amperes. Find:

(a) the phase difference (i.e. the angle ϕ in Fig. 14.14) between the phasors for V and I
(b) the power, given that power $= |V| \times |I| \times \cos\phi$ watts.

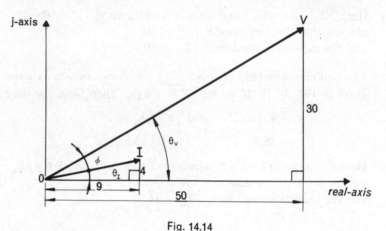

Fig. 14.14

Fig. 14.14 shows a sketch of the Argand diagram showing the phasors for I and V. Phasors in electrical work are usually shown with arrows.

To find $V = 50+j30$ in polar form:

$$|V| = \sqrt{50^2+30^2} \quad \text{and} \quad \tan\theta_V = \frac{30}{50}$$

$$= 58.3 \qquad \therefore \qquad \theta_V = 30° \, 58'$$

To find $I = 9+j4$ in polar form:

$$|I| = \sqrt{9^2+4^2} \quad \text{and} \quad \tan\theta_I = \frac{4}{9}$$

$$= 9.8 \qquad \therefore \qquad \theta_I = 23° \, 58'$$

(a) The phase difference $\phi = \theta_V - \theta_I$

$$= 30° \, 58' - 23° \, 58'$$

$$= 7°$$

(b) $$\text{power} = |V| \times |I| \times \cos\phi$$

$$= 58.3 \times 9.8 \times \cos 7°$$

$$= 567 \text{ watts}$$

Exercise 37

1) Show, indicating each one clearly, the following complex numbers on a single Argand diagram: $4+j3$, $-2+j$, $3-j4$, $-3.5-j2$, j3 and $-j4$.

2) Find the moduli and arguments of the complex numbers $3+j4$ and $4-j3$.

3) If the complex number $z_1 = -3+j2$ find $|z_1|$ and arg z_1.

4) If the complex number $z_2 = -4-j2$ find $|z_2|$ and arg z_2.

5) Express each of the following complex numbers in polar form:

(a) $4+j3$ (b) $3-j4$ (c) $-3+j3$ (d) $-2-j$ (e) j4
(f) $-j3.5$.

6) Convert the following complex numbers, which are given in polar form, into Cartestian form:

(a) $3\underline{/45°}$ (b) $5\underline{/154°}$ (c) $4.6\underline{/-20°}$ (d) $3.2\underline{/-120°}$

7) Simplify the following products of two complex numbers, given in polar form, expressing the answer in polar form:

(a) $8\underline{/30°}\times7\underline{/40°}$ (b) $2\underline{/-20°}\times5\underline{/-30°}$ (c) $5\underline{/120°}\times3\underline{/-30°}$
(d) $7\underline{/-50°}\times3\underline{/-40°}$

8) Simplify the following divisions of two complex numbers, given in polar form, expressing the answer in polar form:

(a) $\dfrac{8\underline{/20°}}{3\underline{/50°}}$ (b) $\dfrac{10\underline{/-40°}}{5\underline{/20°}}$ (c) $\dfrac{3\underline{/-15°}}{5\underline{/-6°}}$ (d) $\dfrac{1.7\underline{/35° \ 17'}}{0.6\underline{/-9° \ 22'}}$

9) Three complex numbers z_1, z_2 and z_3 are given in polar form by $z_1 = 3\underline{/35°}$, $z_2 = 5\underline{/28°}$ and $z_3 = 2\underline{/-50°}$. Simplify:

(a) $z_1\times z_2\times z_3$ giving the answer in polar form.

(b) $\dfrac{z_1\times z_2}{z_3}$ giving the answer in algebraic form.

10) If the complex number $z = 2-j3$ express in polar form:

(a) $\dfrac{1}{z}$ (b) z^2

11) The admittance Y of a circuit is given by $Y = \dfrac{1}{Z}$.

(a) If $Z = 3+j5$ find Y in polar form.
(b) If $Z = 17.4\underline{/42°}$ find Y in algebraic form.

12) Using the notation and information given in the data for the worked Example 7 of the text (p. 222), find the resistance and the inductive or capacitive reactance for each of the following impedances:

(a) $4.5 + 2.2j$ (b) $23 - 35j$ (c) $29.6 \underline{/23° 22'}$ (d) $7 \underline{/-12°}$

13) The potential difference across a circuit is given by the complex number $V = 40 + j35$ volts and the current is given by the complex number $I = 6 + j3$ amperes.   Sketch the appropriate phasors on an Argand diagram and find:

(a) the phase difference (i.e. the angle ϕ) between the phasors for V and I,
(b) the power, given that   power $= |V| \times |I| \times \cos \phi$.

SUMMARY

$$j = \sqrt{-1} \quad \text{or} \quad j^2 = -1$$

The algebraic form of a complex number is $x + jy$.

The polar form of a complex number is $r(\cos \theta + j \sin \theta)$ or $r \underline{/\theta}$.

The conjugate of $(x + jy)$ is $(x - jy)$.

The conjugate of $(x - jy)$ is $(x + jy)$.

If a complex number is multiplied by its conjugate the result is a real number.

The effect of j as an operator on a phasor representing a complex number is to change its direction by 90° anti-clockwise, without any alteration in its magnitude.

If a complex number z in its algebraic form is $z = x + jy$,

and in its polar form is $z = r(\cos \theta + j \sin \theta)$ or $r \underline{/\theta}$

then $\text{mod } z = |z| = r = \sqrt{x^2 + y^2}$

and $\arg z = \theta$ where $\tan \theta = \dfrac{y}{x}$

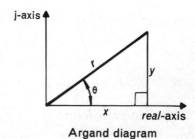

Argand diagram

To multiply two complex numbers in polar form, we multiply their moduli and add their arguments. Thus $(r_1 \underline{/\theta_1}) \times (r_2 \underline{/\theta_2}) = (r_1 \times r_2) \underline{/\theta_1 + \theta_2}$

To divide two complex numbers in polar form, we divide their moduli and subtract their arguments. Thus: $\dfrac{r_1 \underline{/\theta_1}}{r_2 \underline{/\theta_2}} = \left(\dfrac{r_1}{r_2}\right) \underline{/\theta_1 - \theta_2}$

15. APPLICATIONS OF DIFFERENTIATION

After reaching the end of this chapter you should be able to:-

1. *Determine the second derivative $\dfrac{d^2y}{dx^2}$, or f"(x), and higher derivatives of a function.*
2. *Evaluate a second or higher derivative at a given point.*
3. *Calculate linear and angular velocity and acceleration at a given time using first and second derivatives.*
4. *Define a stationary point on a graph.*
5. *Determine the nature of a stationary point by considering the gradient on either side of the point.*
6. *Determine the nature of a stationary point using the sign of a second or higher derivative.*
7. *Solve problems involving maximum and minimum points.*
8. *Sketch curves.*

SECOND AND HIGHER DERIVATIVES

If
$$y = x^6$$

then
$$\frac{dy}{dx} = 6x^5$$

and if we differentiate this equation again with respect to x we obtain

$$\frac{d}{dx}\left(\frac{dy}{dx}\right) = \frac{d}{dx}(6x^5)$$

or
$$\frac{d^2y}{dx^2} = 30x^4$$

Differentiating again gives

$$\frac{d}{dx}\left(\frac{d^2y}{dx^2}\right) = \frac{d}{dx}(30x^4)$$

or
$$\frac{d^3y}{dx^3} = 120x^3$$

and for a fourth time

$$\frac{d^4y}{dx^4} = 360x^2, \quad \text{and so on.}$$

Just as $\dfrac{dy}{dx} = 6x^5$ is called the first derived function or first differential coefficient or first derivative of y with respect to x, so $\dfrac{d^2y}{dx^2} = 30x^4$ is called the second derived function or second derivative of y with respect to x and so on. It should be noted that the 2 which occurs twice in $\dfrac{d^2y}{dx^2}$ is *not* an index but merely indicates that the original function has been differentiated that number of times.

Hence $\dfrac{d^2y}{dx^2}$ is *not* the same as $\left(\dfrac{dy}{dx}\right)^2$.

The optional function notation is: $\quad y = f(x)$

$$\frac{dy}{dx} = f'(x)$$

$$\frac{d^2y}{dx^2} = f''(x)$$

$$\frac{d^3y}{dx^3} = f'''(x), \quad \text{and so on.}$$

EXAMPLE 1

Evaluate $\dfrac{dy}{dx}$ and $\dfrac{d^2y}{dx^2}$ when $\quad x = 2 \quad$ given that $\quad y = \sqrt{x} + \sin x$.

Now
$$y = \sqrt{x} + \sin x$$
$$= x^{1/2} + \sin x$$

$\therefore \qquad\qquad \dfrac{dy}{dx} = \tfrac{1}{2}x^{-1/2} + \cos x$

$\therefore \qquad\qquad \dfrac{d^2y}{dx^2} = -\dfrac{1}{4}x^{-3/2} - \sin x$

Before substituting the numerical value of x it is recommended that the expressions for $\dfrac{dy}{dx}$ and $\dfrac{d^2y}{dx^2}$ are transformed to give positive indices.

Although most electronic calculators will evaluate expressions with negative indices, mistakes may occur and it is more difficult to detect a computational error.

Remember also that sin 2 is the sine of 2 radians, *not* of 2 degrees, and that your calculator should be in radian mode.

Now
$$\frac{dy}{dx} = \frac{1}{2x^{1/2}} + \cos x = \frac{1}{2\sqrt{x}} + \cos x$$

Hence when $x = 2$, $\dfrac{dy}{dx} = \dfrac{1}{2\sqrt{2}} + \cos 2 = 0.353\,6 + (-0.416\,1)$

$$= -0.062\,5$$

Also $\dfrac{d^2y}{dx^2} = -\dfrac{1}{4x^{3/2}} - \sin x = -\dfrac{1}{4(\sqrt{x})^3} - \sin x$

and when $x = 2$,

$$\dfrac{d^2y}{dx^2} = -\dfrac{1}{4(\sqrt{2})^3} - \sin 2 = -0.088\,4 - 0.909\,3$$

$$= -0.997\,7$$

EXAMPLE 2

Given that $v = 4 \sin\left(100\pi t + \dfrac{\pi}{3}\right)$ calculate the value of $\dfrac{dv}{dt}$ when $t = 0.001$.

Now $v = 4 \sin\left(100\pi t + \dfrac{\pi}{3}\right)$

Hence $\dfrac{dv}{dt} = 4 \cos\left(100\pi t + \dfrac{\pi}{3}\right) \times 100\pi$ (function of a function)

$$= 400\pi \cos\left(100\pi t + \dfrac{\pi}{3}\right)$$

When $t = 0.001$; $\dfrac{dv}{dt} = 400\pi \cos(1.36)$ ... (angle in radians)

$$\approx 263$$

Exercise 38

1) If $y = 3x^3 + 2x - 7$ determine an expression for $\dfrac{d^2y}{dx^2}$ and evaluate it when $x = 3$.

2) Evaluate $f'(x)$ and $f''(x)$ when $x = -2$ given that $y = 5x^4 + 7x^2 + x$.

3) Given that $\theta = 18 + 20t^2 - 2t^3$ find the value of $\dfrac{d^2\theta}{dt^2}$ when $t = 5$.

4) If $i = -2 \sin(20t - 0.02)$ evaluate $\dfrac{d^2i}{dt^2}$ when $t = 0.01$ (radian measure).

5) If $Z = 5 \cos 3\theta$ find the value of $\dfrac{dZ}{d\theta}$, $\dfrac{d^2Z}{d\theta^2}$ and $\dfrac{d^3Z}{d\theta^3}$ when $\theta = \dfrac{\pi}{2}$.

6) Determine the value of $\dfrac{d^2p}{d\phi^2}$ given that $p = 6 \sin 4\phi$, where ϕ is measured in radians and has a value equivalent to 28°.

7) If $f(\alpha) = 2 \cos \dfrac{\alpha}{4}$ determine the value of $f''(\alpha)$ when $\alpha = 1.6$ radians.

8) Calculate the value of $\dfrac{d^2v}{dt^2}$ when $t = 0.002$ and

$$v = -5 \sin\left(100\pi t - \dfrac{\pi}{6}\right).$$

VELOCITY AND ACCELERATION

Suppose that a vehicle starts from rest and travels 60 metres in 12 seconds. The average velocity may be found by dividing the total distance travelled by the total time taken, that is $\dfrac{60}{12} = 5$ m/s. This is NOT the INSTANTANEOUS velocity, however, AT a time of 12 seconds, but is the AVERAGE VELOCITY over the 12 seconds as calculated previously.

Fig. 15.1 shows a graph of distance s, against time t. The average velocity over a period is given by the gradient of the chord which meets the curve at the extremes of the period. Thus in the diagram the gradient of the dotted chord QR gives the average velocity between $t = 2$ s and $t = 6$ s. It is found to be $\frac{13}{4} = 3.25$ m/s.

The velocity at any point is the rate of change of s with respect t and may be found by finding the gradient of the curve at that point. In mathematical notation this is given by $\dfrac{ds}{dt}$.

Suppose we know that the relationship between s and t is

$$s = 0.417t^2$$

then velocity, $\qquad v = \dfrac{ds}{dt} = 0.834t$

and hence when $t = 12$ seconds, then $v = 0.834 \times 12 = 10$ m/s.

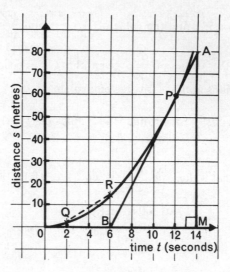

Fig. 15.1

This result may be found graphically by drawing the tangent to the curve of s against t at the point P and constructing a suitable right angled triangle ABM.

Hence the velocity at $\quad$ P $= \dfrac{\text{AM}}{\text{BM}} = \dfrac{80}{8} = 10$ m/s $\quad$ which verifies the theoretical result.

Similarly, the rate of change of velocity with respect to time is called acceleration and is given by the gradient of the velocity–time graph at any point. In mathematical notation this is given by $\dfrac{\mathrm{d}v}{\mathrm{d}t}$.

Now $$\frac{\mathrm{d}v}{\mathrm{d}t} = \frac{\mathrm{d}}{\mathrm{d}t}(v) = \frac{\mathrm{d}}{\mathrm{d}t}\left(\frac{\mathrm{d}s}{\mathrm{d}t}\right) = \frac{\mathrm{d}^2 s}{\mathrm{d}t^2}$$

and so the acceleration, a, is given by either

$$\frac{\mathrm{d}v}{\mathrm{d}t} \quad \text{or} \quad \frac{\mathrm{d}^2 s}{\mathrm{d}t^2}.$$

The above reasoning was applied to linear motion, but it could also have been used for angular motion. The essential difference is that distance, s, is replaced by angle turned through, θ rad.

Both sets of results are summarised in Fig. 15.2.

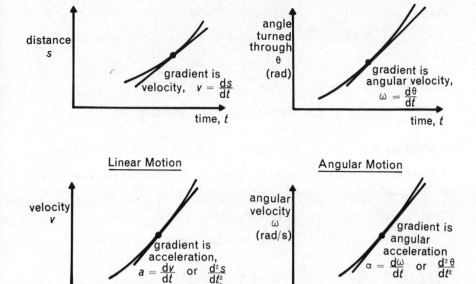

Fig. 15.2

EXAMPLE 3

A body moves a distance s metres in a time of t seconds so that $s = 2t^3 - 9t^2 + 12t + 6$.

Find (a)   its velocity after 3 seconds,

(b)   its acceleration after 3 seconds, and

(c)   when the velocity is zero.

We have,
$$s = 2t^3 - 9t^2 + 12t + 6$$

$\therefore$
$$\frac{ds}{dt} = 6t^2 - 18t + 12$$

and
$$\frac{d^2s}{dt^2} = 12t - 18$$

(a)   When $t = 3$ then velocity is $\dfrac{ds}{dt} = 6(3)^2 - 18(3) + 12 = 12$ m/s

(b)   when $t = 3$ then acceleration is $\dfrac{d^2s}{dt^2} = 12(3) - 18 = 18$ m/s²

(c)   when the velocity is zero then $\dfrac{ds}{dt} = 0$

that is $$6t^2-18t+12 = 0$$

$\therefore$ $$t^2-3t+2 = 0$$

$\therefore$ $$(t-1)(t-2) = 0$$

$\therefore$ either $\qquad t-1 = 0 \quad$ or $\quad t-2 = 0$

$\therefore$ either $\qquad t = 1$ second or $t = 2$ seconds.

EXAMPLE 4

The angle θ radians is connected with the time t seconds by the relationship $\theta = 20+5t^2-t^3$.

Find (a)   the angular velocity when $t = 2$ seconds,   and

(b)   the value of t when the angular deceleration is 4 rad/s².

We have, $$\theta = 20+5t^2-t^3$$

$\therefore$ $$\frac{d\theta}{dt} = 10t-3t^2$$

and $$\frac{d^2\theta}{dt^2} = 10-6t$$

(a)   when $t = 2$,   then the angular velocity $\dfrac{d\theta}{dt} = 10(2)-3(2)^2 = 8$ rad/s.

(b)   an angular deceleration of 4 rad/s² may be called an angular acceleration of -4 rad/s²,

$\therefore$ when $\qquad \dfrac{d^2\theta}{dt^2} = -4 \quad$ then $\quad -4 = 10-6t$

$$\text{or} \qquad t = 2.33 \text{ seconds.}$$

Exercise 39

1)   If $s = 10+50t-2t^2$,   where s metres is the distance travelled in t seconds by a body, what is the velocity of the body after 2 seconds?

2)   If $v = 5+24t-3t^2$ where v m/s is the velocity of a body at a time t seconds, what is the acceleration when $t = 3$?

3)   A body moves s metres in t seconds where $s = t^3-3t^2-3t+8$.

Find   (a)   its velocity at the end of 3 seconds,

(b)   when its velocity is zero,

(c)   its acceleration at the end of 2 seconds,

(d)   when its acceleration is zero.

4) A body moves s metres in t seconds where $s = \dfrac{1}{t^2}$. Find the velocity and acceleration after 3 seconds.

5) The distance s metres travelled by a falling body starting from rest after a time t seconds is given by $s = 5t^2$. Find its velocity after 1 second and after 3 seconds.

6) The distance s metres moved by the end of a lever after a time t seconds is given by the formula $s = 6t^2$. Find the velocity of the end of the lever when it has moved a distance $\frac{1}{2}$ metre.

7) The angular displacement θ radians of the spoke of a wheel is given by the expression $\theta = \dfrac{1}{2}t^4 - t^3$ where t seconds is the time.

Find   (a)   the angular velocity after 2 seconds,

       (b)   the angular acceleration after 3 seconds,

       (c)   when the angular acceleration is zero.

8) An angular displacement θ radians in time t seconds is given by the equation $\theta = \sin 3t$.

Find   (a)   the angular velocity when $t = 1$ second,

       (b)   the smallest positive value of t for which the angular velocity is 2 rad/s,

       (c)   the angular acceleration when $t = 0.5$ seconds,

       (d)   the smallest positive value of t for which the angular acceleration is 9 rad/s².

9) A mass of 5000 kg moves along a straight line so that the distance s metres travelled in a time t seconds is given by $s = 3t^2 + 2t + 3$. If v m/s is its velocity and m kg is its mass then its kinetic energy is given by the formula $\dfrac{1}{2}mv^2$. Find its kinetic energy at a time $t = 0.5$ seconds remembering that the joule (J) is the unit of energy.

CRITICAL POINTS

Any point on the graph of $y = f(x)$ such that $f'(x) = 0$, or $\dfrac{\mathrm{d}y}{\mathrm{d}x} = 0$, is called a *critical point* on the curve. At this point the tangent is parallel to the x-axis and for this reason the term *stationary point* is often used instead of critical point.

The nature of the curve in the neighbourhood of a critical point is important.

It is possible at this point to have a *local maximum turning point*, or a *local minimum turning point* or a *local point of horizontal inflexion*.

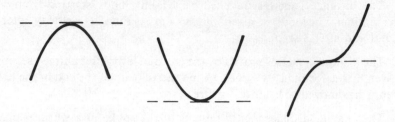

Maximum and minimum turning points and problems will be investigated first and then the theory will be expanded to include points of inflexion and curve sketching.

TURNING POINTS

At the points P and Q (Fig. 15.3) the tangent to the curve is parallel to the *x*-axis. The points P and Q are called *turning points*. The turning point at P is called a *maximum* turning point and the turning point at Q is called a *minimum* turning point. It will be seen from Fig. 15.3 that the value of y at P is not the greatest value of y nor is the value of y at Q the least. The terms maximum and minimum values apply only to the values of y at the turning points and not to the values of y in general.

In practical applications, however, we are usually concerned with a specific range of values of x which are dictated by the problem. There is then no difficulty in identifying a particular maximum or minimum within this range of values of x.

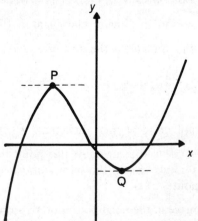

Fig. 15.3

EXAMPLE 5

Plot the graph of $y = x^3 - 5x^2 + 2x + 8$ for values of x between -2 and 6. Hence find the maximum and minimum values of y.

To plot the graph we draw up a table in the usual way.

x	-2	-1	0	1	2	3	4	5	6
$y = x^3 - 5x^2 + 2x + 8$	-24	0	8	6	0	-4	0	18	56

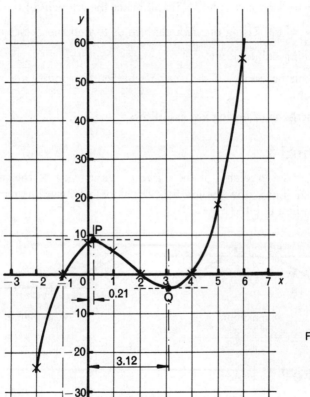

Fig. 15.4

The graph is shown in Fig. 15.4. The maximum value occurs at the point **P** where the tangent to the curve is parallel to the x-axis. The minimum value occurs at the point **Q** where again the tangent to the curve is parallel to the x-axis. From the graph the maximum value of y is 8.21 and the minimum value of y is -4.06.

Notice that the value of y at **P** is not the greatest value of y nor is the value of y at **Q** the least. However, the values of y at **P** and **Q** are called the maximum and minimum values of y respectively.

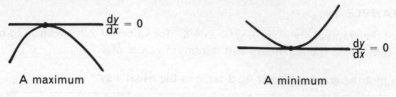

Fig. 15.5

It is not always convenient to draw the full graph to find the turning points as in the previous example. At a turning point the tangent to the curve is parallel to the x-axis (Fig. 15.5) and hence the gradient of the curve is zero, i.e. $\dfrac{dy}{dx} = 0$. Using this fact enables us to find the values of x at which the turning points occur.

It is then necessary to determine whether the points are maximum or minimum.

Two methods of testing are as follows:

Method 1

Consider the gradients of the curve on either side of the turning point. Fig. 15.6 shows how the gradient (or slope) of curve changes the vicinity of a turning point.

Fig. 15.6

Method 2

Find the value of $\dfrac{d^2y}{dx^2}$ at the turning point.

If it is positive then the turning point is a minimum, and if it is negative then the turning point is a maximum.

If the original expression may be differentiated twice and the expression for $\dfrac{d^2y}{dx^2}$ obtained without too much difficulty, then the second method is generally used.

EXAMPLE 6

Find the maximum and minimum values of y given that
$y = x^3 + 3x^2 - 9x + 6$.

We have
$$y = x^3 + 3x^2 - 9x + 6$$

$\therefore$
$$\frac{dy}{dx} = 3x^2 + 6x - 9$$

and
$$\frac{d^2y}{dx^2} = 6x + 6$$

At a turning point
$$\frac{dy}{dx} = 0$$

$\therefore$ $3x^2 + 6x - 9 = 0$

$\therefore$ $x^2 + 2x - 3 = 0$ by dividing through by 3.

$\therefore$ $(x-1)(x+3) = 0$

$\therefore$ either $x - 1 = 0$ or $x + 3 = 0$

$\therefore$ either $x = 1$ or $x = -3$

Test for maximum or minimum:

From above we have $\dfrac{d^2y}{dx^2} = 6x + 6$

$\therefore$ at the point where $x = 1$, $\dfrac{d^2y}{dx^2} = 6(1) + 6 = +12$

This is positive and hence the turning point at $x = 1$ is a minimum.

The minimum value of y may be found by substituting $x = 1$ into the given equation. Hence:

$$y_{min} = (1)^3 + 3(1)^2 - 9(1) + 6 = 1.$$

At the point where $x = -3$, $\dfrac{d^2y}{dx^2} = 6(-3) + 6 = -12.$

This is negative and hence at $x = -3$ there is a maximum turning point. The maximum value of y may be found by substituting $x = -3$ into the given equation. Hence:

$$y_{max} = (-3)^3 + 3(-3)^2 - 9(-3) + 6 = +33.$$

To illustrate the test for maximum and minimum using the tangent gradient method, this method will be used to verify the above results.

At the turning point where $x = 1$,   we know that

$$\frac{dy}{dx} = 0, \quad \text{i.e. there is zero slope,}$$

and using a value of x slightly less than 1, say $x = 0.5$ then

$$\frac{dy}{dx} = 3(0.5)^2 + 6(0.5) - 9 = -5.25 \quad \text{i.e. there is a negative slope,}$$

and using a value of x slightly greater than 1, say $x = 1.5$ then

$$\frac{dy}{dx} = 3(1.5)^2 + 6(1.5) - 9 = +6.75 \quad \text{i.e. there is a positive slope.}$$

These results are best shown by means of a diagram (Fig. 15.7) which indicates clearly that when $x = 1$ we have a minimum.

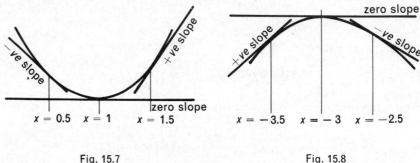

Fig. 15.7   Fig. 15.8

Now at the turning point where $x = -3$ we know that

$$\frac{dy}{dx} = 0, \quad \text{i.e. there is zero slope,}$$

and using a value of x slightly less than -3, say $x = -3.5$ then

$$\frac{dy}{dx} = 3(-3.5)^2 + 6(-3.5) - 9 = +6.75 \quad \text{i.e. there is a positive slope,}$$

and using a value of x slightly greater than -3, say $x = -2.5$ then

$$\frac{dy}{dx} = 3(-2.5)^2 + 6(-2.5) - 9 = -5.25 \quad \text{i.e. there is a negative slope.}$$

Fig. 15.8 indicates that when $x = -3$ we have a maximum turning point.

There are many applications in engineering which involve the finding of maxima and minima. The first step is to construct an equation connecting the quantity for which a maximum or minimum is required in terms of another variable. A diagram representing the problem may help in the formation of this initial equation.

EXAMPLE 7

A rectangular sheet of metal 360 mm by 240 mm has four equal squares cut out at the corners. The sides are then turned up to form a rectangular box. Find the length of the sides of the squares cut out so that the volume of the box may be as great as possible, and find this maximum volume.

Let the length of the side of each cut away square be x mm as shown in Fig. 15.9.

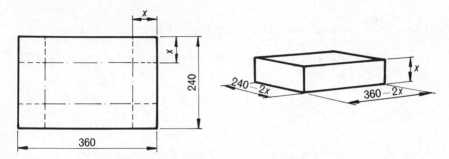

Fig. 15.9

Hence the volume is

$$V = x(240-2x)(360-2x)$$

$$= 4x^3 - 1200x^2 + 86\ 400x$$

$\therefore$

$$\frac{\mathrm{d}V}{\mathrm{d}x} = 12x^2 - 2400x + 86\ 400$$

and

$$\frac{\mathrm{d}^2V}{\mathrm{d}x^2} = 24x - 2400$$

At a turning point

$$\frac{\mathrm{d}V}{\mathrm{d}x} = 0$$

$\therefore$ $$12x^2 - 2400x + 86\ 400 = 0$$

or $$x^2 - 200x + 7200 = 0 \quad \text{by dividing through by 12.}$$

Now this is a quadratic equation which does not factorise so we will have to solve using the formula for the standard quadratic $ax^2 + bx + c = 0$ which gives $x = \dfrac{-b \pm \sqrt{b^2 - 4ac}}{2a}$. Hence the solution of our equation

is
$$x = \frac{-(-20)\pm\sqrt{(-20)^2-4\times1\times72}}{2\times1}$$

∴ either $x = 152.9$ or $x = 47.1$

However, from the physical sizes of the sheet, it is not possible for x to be 152.9 mm (since one side is only 240 mm long) so we reject this solution. Hence $x = 47.1$ mm.

Test for maximum or minimum:

From the above we have $\dfrac{\mathrm{d}^2V}{\mathrm{d}x^2} = 24x-2400$

and hence when $x = 47.1$, $\dfrac{\mathrm{d}^2V}{\mathrm{d}x^2} = 24(47.1)-2400 = -1270$

This is negative and hence V is a maximum when $x = 47.1$ mm.

It only remains to find the maximum volume by substituting $x = 47.1$ into the equation for V.

∴ $V_{max} = 47.1(240-2\times47.1)(360-2\times47.1) = 1.825\times10^6$ mm³.

EXAMPLE 8

A cylinder with an open top has a capacity of 2 m³ and is made from sheet metal. Neglecting any overlaps at the joints find the dimensions of the cylinder so that the amount of sheet steel used is a minimum.

Let the height of the cylinder be h metres and the radius of the base be r metres as shown in Fig. 15.10.

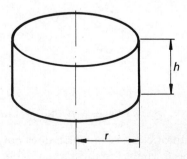

Fig. 15.10

Now the total area of metal = area of base+area of curved side:

$$\therefore \qquad A = \pi r^2 + 2\pi rh$$

We cannot proceed to differentiate as there are two variables on the right hand side of the equation. It is possible, however, to find a connection between r and h using the fact that the volume is 2 m².

Now $\qquad\qquad$ volume of a cylinder $= \pi r^2 h$

$$\therefore \qquad\qquad\qquad 2 = \pi r^2 h$$

from which $\qquad\qquad\qquad h = \dfrac{2}{\pi r^2}$

We may now substitute for h is the equation for A,

$$\therefore \qquad\qquad A = \pi r^2 + 2\pi r\left(\frac{2}{\pi r^2}\right)$$

$$\therefore \qquad\qquad\quad = \pi r^2 + \frac{4}{r}$$

$$\therefore \qquad\qquad\quad = \pi r^2 + 4r^{-1}$$

$$\therefore \qquad\qquad \frac{\mathrm{d}A}{\mathrm{d}r} = 2\pi r - 4r^{-2}$$

and $\qquad\qquad \dfrac{\mathrm{d}^2 A}{\mathrm{d}r^2} = 2\pi + 8r^{-3}$

Now for a turning point $\qquad \dfrac{\mathrm{d}A}{\mathrm{d}r} = 0$

or $\qquad\qquad 2\pi r - 4r^{-2} = 0$

$$\therefore \qquad\qquad 2\pi r - \frac{4}{r^2} = 0$$

$$\therefore \qquad\qquad\quad 2\pi r = \frac{4}{r^2}$$

$$\therefore \qquad\qquad\quad r^3 = \frac{2}{\pi} = 0.637$$

$$r = \sqrt[3]{0.637} = 0.860$$

To test for a minimum:

From above we have $\qquad \dfrac{\mathrm{d}^2 A}{\mathrm{d}r^2} = 2\pi + 8r^{-3}$

$$= 2\pi + \frac{8}{r^3}$$

We do not need to do any further calculation here as this expression must be positive for all positive values of r. Hence $r = 0.86$ makes A a minimum.

We may find the corresponding value of h by substituting $r = 0.86$ into the equation found previously for h in terms of r

$$\therefore \qquad h = \frac{2}{\pi(0.86)^2} = 0.86$$

hence for the minimum amount of metal to be used the radius is 0.86 m and the height is 0.86 m.

Exercise 40

1) Find the maximum and minimum values of:

(a) $y = 2x^3 - 3x^2 - 12x + 4$

(b) $y = x^3 - 3x^2 + 4$

(c) $y = 6x^2 + x^3$

2) Given that $y = 60x + 3x^2 - 4x^3$, calculate:

(a) the gradient of the tangent to the curve of y at the point where $x = 1$;

(b) the value of x for which y has its maximum value;

(c) the value of x for which y has its minimum value.

3) Calculate the co-ordinates of the points on the curve $y = x^3 - 3x^2 - 9x + 12$ at each of which the tangent to the curve is parallel to the x-axis.

4) A curve has the equation $y = 8 + 2x - x^2$. Find:

(a) the value of x for which the gradient of the curve is 6;

(b) the value of x which gives the maximum value of y;

(c) the maximum value of y.

5) The curve $y = 2x^2 + \dfrac{k}{x}$ has a gradient of 5 when $x = 2$.

Calculate (a) the value of k; (b) the minimum value of y.

6) From a rectangular sheet of metal measuring 120 mm by 75 mm equal squares of side x are cut from each of the corners. The remaining flaps are then folded upwards to form an open box. Prove that the volume of the box is given by $V = 9000x - 390x^2 + 4x^3$. Find the value of x such that the volume is a maximum.

7) An open rectangular tank of height h metres with a square base of side x metres is to be constructed so that it has a capacity of 500 cubic metres. Prove that the surface area of the four walls and the base will be $\left(\dfrac{2000}{x}+x^2\right)$ square metres. Find the value of x for this expression to be a minimum.

8) The volume of a cone is given by the formula $V=\frac{1}{3}\pi r^2 h$, where h is the height of the cone and r its radius. If $h=6-r$, calculate the value of r for which the volume is a maximum.

9) A box without a lid has a square base of side x mm and rectangular sides of height h mm. It is made from 10 800 mm² of sheet metal of negligible thickness. Prove that $h=\dfrac{10\,800-x^2}{4x}$ and that the volume of the box is $(2700x-\frac{1}{4}x^3)$. Hence calculate the maximum volume of the box.

10) A cylindrical tank, with an open top, is to be made to hold 300 cubic metres of liquid. Find the dimensions of the tank so that its surface area shall be a minimum.

11) A cooling tank is to be made with the trapezoidal section as shown:

Its cross-sectional area is to be 300 000 mm². Show that the width of material needed to form, from one sheet, the bottom and folded-up sides is $w=\dfrac{300\,000}{h}+1.828h$. Hence find the height h of the tank so that the width of material needed is a minimum.

12) A cylindrical cup is to be drawn from a disc of metal of 50 mm diameter. Assuming that the surface area of the cup is the same as that of the disc find the dimensions of the cup so that its volume is a maximum.

13) A lever weighing 12 N per m run of its length is as shown:

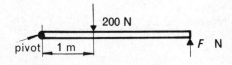

Find the length of the lever so that the force F shall be a minimum.

CURVE SKETCHING

Accurately drawn graphs are very important in engineering, but it is also useful to make a rough sketch of a curve without having to calculate and plot a large number of points. Once the position and nature of the critical points on a simple curve have been determined, a rough outline can usually be sketched.

When using the procedure, Method 2 of the previous section, that if the value of $\dfrac{d^2y}{dx^2}$ at a point is positive we have a minimum and if negative a maximum, it is possible to have $\dfrac{d^2y}{dx^2} = 0$; that is neither positive nor negative.

When $\dfrac{d^2y}{dx^2}$ 'vanishes' like this, but differentiation of the derived function still remains feasible, the following rules can be applied. In each case the critical point being considered is a maximum or minimum turning point or a horizontal point of inflexion.

1) If a critical point occurs when $x = a$, then $\dfrac{dy}{dx} = 0$ when $x = a$.

2) For a *maximum* turning point the first 'non-vanishing' derivative at $x = a$ must be of *even order and negative*.

$\dfrac{d^2y}{dx^2}, \dfrac{d^4y}{dx^4}, \dfrac{d^6y}{dx^6},$ are even order derivatives.

3) For a *minimum* turning point the first non-vanishing derivative at $x = a$ must be of *even order and positive*.

4) For a *horizontal point of inflexion* the first non-vanishing derivative at $x = a$ must be of *odd order*.

$\dfrac{d^3y}{dx^3}, \dfrac{d^5y}{dx^5},$ are odd order derivatives.

EXAMPLE 9

Sketch the curve represented by $y = x^4$.

Now $$y = x^4$$

$\therefore$ $$\frac{dy}{dx} = 4x^3$$

At a turning point $\qquad \dfrac{dy}{dx} = 0$

$\therefore \qquad\qquad\qquad\qquad 4x^3 = 0$

$\therefore \qquad\qquad\qquad\qquad\; x = 0$

Thus only one turning point exists at the point where $x = 0$. To determine its nature we use

$$\dfrac{d^2y}{dx^2} = 12x^2$$

This is of *even order* but it 'vanishes' when $x = 0$. That is it is neither positive nor negative.

Again $\qquad\qquad\qquad \dfrac{d^3y}{dx^3} = 24x$

This is of *odd order* but again it vanishes when $x = 0$.

Again $\qquad\qquad\qquad \dfrac{d^4y}{dx^4} = 24$

This does not vanish when $x = 0$ (since it does not contain x) and is of *even order* and *positive*. Hence when $x = 0$ we have a minimum turning point. Substituting $x = 0$ into $y = x^4$ fixes the position of the turning point at $(0, 0)$.

A sketch is shown in Fig. 15.11.

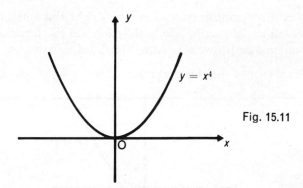

$y = x^4$

Fig. 15.11

EXAMPLE 10

Determine the nature and position of any turning points on the curve represented by $y = x^3$ and hence sketch the curve.

Now $$y = x^3$$

$\therefore$ $$\frac{dy}{dx} = 3x^2$$

At a critical point $$\frac{dy}{dx} = 0$$

$\therefore$ $$3x^2 = 0$$

$\therefore$ $$x = 0$$

Thus there is only one turning point where $x = 0$. To determine its nature we use

$$\frac{d^2y}{dx^2} = 6x$$

This vanishes when $x = 0$, and

$$\frac{d^3y}{dx^3} = 6.$$

This does not vanish when $x = 0$, and is of *odd order*, hence the stationary point is a horizontal point of inflexion.

Substituting $x = 0$ into $y = x^3$ fixes the point of inflexion at the origin $(0, 0)$.

We now have two possible sketches; a curve rising from bottom left to top right of the paper with inflection at the origin or a similar graph falling from the top left to the bottom right of the paper.

Just as a positive gradient rises from left to right and a negative gradient falls from left to right so the line of inflexion will rise from left to right if the odd ordered derivative is positive and fall if negative.

Hence the sketch of the curve represented by $y = x^3$ is shown in Fig. 15.12.

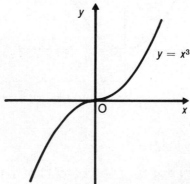

Fig. 15.12

EXAMPLE 11

Determine and identify the stationary points of the curve represented by

$$y = 2x^4 - 8x^3 + 1$$

and then sketch the graph.

Now
$$y = 2x^4 - 8x^3 + 1$$

∴
$$\frac{dy}{dx} = 8x^3 - 24x^2$$

At a critical point
$$\frac{dy}{dx} = 0$$

∴
$$8x^3 - 24x^2 = 0$$

∴
$$8x^2(x - 3) = 0$$

∴
$$x = 0 \quad \text{or} \quad x = 3$$

To investigate the stationary point at $x = 0$ we have

$$\frac{d^2y}{dx^2} = 24x^2 - 48x$$

which vanishes when $x = 0$.

Again
$$\frac{d^3y}{dx^3} = 48x - 48$$

which does not vanish when $x = 0$ and is of *odd order* and *negative* at this point.

Substituting $x = 0$ into the equation $y = 2x^4 - 8x^3 + 1$ now gives us the information that at the point $(0, 1)$ we have a horizontal point of inflexion falling from left to right.

To investigate the stationary point at $x = 3$ we have

$$\frac{d^2y}{dx^2} = 24(3^2) - 48(3) = 72$$

Also
$$y = 2(3^4) - 8(3^3) + 1 = -53$$

Hence at the point $(3, -53)$ we have a minimum turning point. See Fig. 15.13.

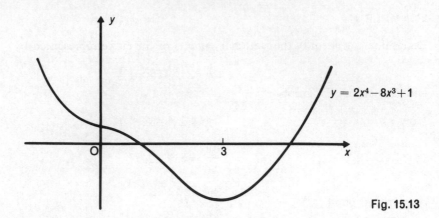

$y = 2x^4 - 8x^3 + 1$

Fig. 15.13

EXAMPLE 12

Show that the function $y = 6\cos\theta + 8\sin\theta$ has stationary points where $\tan\theta = \frac{4}{3}$. Determine the nature and coordinates of the two stationary points in the range $0 \leqslant \theta \leqslant 2\pi$ and then sketch this part of the graph.

Now
$$y = 6\cos\theta + 8\sin\theta$$

$\therefore$
$$\frac{dy}{d\theta} = -6\sin\theta + 8\cos\theta$$

At a stationary point $\dfrac{dy}{d\theta} = 0$

$\therefore$
$$-6\sin\theta + 8\cos\theta = 0$$

$$\sin\theta = \frac{8}{6}\cos\theta$$

or
$$\tan\theta = \frac{4}{3}$$

Thus
$$\theta \, \epsilon \, \{0.9, \, 4.1\}$$

where θ is measured in radians and in the range $0 \leqslant \theta \leqslant 2\pi$.

This shows that we have two stationary points in this range and to determine their nature we use

$$\frac{d^2y}{d\theta^2} = -6\cos\theta - 8\sin\theta$$

Substituting 0.9 and 4.1 into this equation and the original equation indicates that we have a maximum turning point at $(0.9, 10)$ and a minimum turning point at $(4.1, -10)$. See Fig. 15.14.

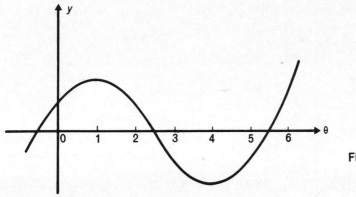

Fig. 15.14

Exercise 41

Use calculus to determine the nature of the critical points on the graphs represented by each of the equations in Questions 1–5 and then sketch each of the curves.

1) $y = 2x^2 + x - 6$

2) $y = x^2(2x + 3)$

3) $y = 2x^3 + 9x^2 - 4$

4) $y = x^3(4 - x)$

5) $y = x^4 - 2x^2 - 2$

6) Show that the function $f(x) = \cos x + \sin x$ has critical points where $\tan x = 1$. Determine the nature and coordinates of the two stationary points in the range $0 \leqslant x \leqslant 2\pi$ and sketch this part of the curve.

7) Sketch the curve represented by $\theta = \cos t + \sqrt{3} \sin t$.

ANSWERS

ANSWERS TO CHAPTER 1

Exercise 1

1) 2, 4   2) 1, 3   3) 2, 5
4) 3, 2   5) 4, 3   6) 3, 4
7) 1.5, 0.2   8) 0.4, -0.9 9) 3, 4
10) 2, 2   11) 5, 8   12) 3, 7
13) 0.3, 0.2, 4.7   14) 25, 0.005, 31.25
15) 100, 0.8, £8100   16) £0.10, £0.30
17) £45, £60   18) $R_1 = 10$, $R_2 = 5$

ANSWERS TO CHAPTER 2

Exercise 2

1) $x^2 - 4x + 3 = 0$
2) $x^2 + 2x - 8 = 0$
3) $x^2 + 3x + 2 = 0$
4) $x^2 - 2.3x + 1.12 = 0$
5) $x^2 - 1.07x - 4.53 = 0$
6) $x^2 + 7.32x + 12.19 = 0$
7) $x^2 - 1.4x = 0$ 8) $x^2 + 4.36x = 0$
9) $x^2 - 12.25 = 0$ 10) $x^2 - 8x + 16 = 0$

Exercise 3

1) ± 6 2) ± 1.25
3) ± 1.333 4) -4 or -5
5) 8 or -9 6) 2 or $\frac{1}{3}$
7) 3   8) 4 or -8
9) $\frac{4}{7}$ or $\frac{3}{2}$ 10) $\frac{7}{3}$ or $-\frac{4}{3}$
11) 1.175 or -0.425 12) $\frac{5}{8}$ or $\frac{1}{6}$
13) 0.573 or -2.907 14) 0.211 or -1.354
15) 1 or -0.2 16) 3.886 or -0.386
17) 0.956 or -1.256 18) 2.388 or 0.262
19) 0.44 or -3.775 20) 8.385 or -2.385
21) -0.225 or -1.775
22) 11.14 or -3.14 23) 1.303 or -2.303
24) ± 53.67 25) 5.24 or 0.76
26) -3.064 or -0.935

Exercise 4

1) 149.55   2) 9.236 cm
3) 6.5 m $\times$ 9.5 m   4) 4 cm

5) 0.685 cm or 23.315 cm
6) 30 mm   7) 50
8) 2.875 cm   9) 54.6 mm
10) 9.464 cm

ANSWERS TO CHAPTER 3

Exercise 5

1) $4.748\,3 \times 10^{-4}$ 2) 0.489 43
3) 66.795 8   4) $1.923\,8 \times 10^{-5}$
5) 2900.6   6) $1.306\,1 \times 10^{-3}$
7) 0.483 74   8) $5.479\,8 \times 10^{-12}$
9) 0.506 85   10) $-3.228\,7$
11) 7.771 3   12) $-9\,209.2$
13) 12.865   14) 1.518 6
15) £143.16   16) 1622
17) 3.666 9

ANSWERS TO CHAPTER 4

Exercise 6

1) a^{11} 2) z^{11} 3) y^{12}
4) 2^8 5) 3^8 6) $\frac{3}{32}a^6$
7) a^3 8) m^7 9) 2^4
10) x^{15} 11) a^4 12) q^8
13) m 14) l^2 15) L^2
16) x^{12} 17) a^{15} 18) $9x^8$
19) 2^6 20) 10^6 21) a^3b^6
22) $a^4b^8c^{12}$ 23) $2^5x^{10}y^{15}z^5$
24) $\dfrac{3^5 m^{10}}{4^5 n^{15}}$ 25) $\frac{1}{10}, \frac{1}{4}, \frac{1}{81}, \frac{1}{25}$
26) 32, 625, 2, 3   27) $3^6, 3^{15}, 3^{12}, 3^{17}$
28) $a^{1/5}, a^{2/3}, a^{4/7}, a^3$ 29) 64, 1 000 000, 4, 8
30) 4, 8, $\frac{1}{3}, \frac{1}{3}$ 31) 10
32) $\frac{5}{8}$ 33) 1, $\frac{1}{5}$, 100 000
34) $\dfrac{1}{3x^2}, \dfrac{1}{3x^2}, \dfrac{1}{5a^4}$ 35) $32a^5$

Exercise 7

1) 3   2) 3   3) 4   4) 3   5) 9
6) 64   7) 100   8) 1   9) 2   10) 3

Exercise 8

1) 2.099 4	**2)** 0.392 77
3) 0.739 28	**4)** 3.453 7
5) 0.330 72	**6)** 0.099 276
7) 0.748 79	**8)** 1.849 2
9) 4.530 1	**10)** 0.071 529
11) 2.084 5	**12)** 368.48
13) 0.795 42	**14)** 0.000 179 16
15) 0.266 99	**16)** 3.594 4
17) 3.059 7	**18)** 0.093 370
19) 3.559 5	**20)** 6.664 6
21) 0.523 33	**22)** 201.18
23) 5.065 3	**24)** 1.199 6

25) 1.773 8   **26)** $-7.379\ 3$ **27)** $-0.322\ 47$
28) 4.627 6   **29)** 2.354 4   **30)** $-2.278\ 5$
31) $-36.536\ 6$ **32)** 22.255 9

Self Test 1

1) d	**2)** a	**3)** b, c	**4)** d
5) a	**6)** c, d	**7)** b	**8)** d
9) a	**10)** a	**11)** a	**12)** c
13) d	**14)** c		

ANSWERS TO CHAPTER 5

Exercise 9

1) (a) 2.095   (b) 1.957   (c) 3.003
    (d) 5.741   (e) -0.711 (f) -4.806
2) (a) 3.584   (b) 6.178   (c) -0.0456
    (d) -3.267 (e) 0.728   (f) -2.303
3) (a) 15.80   (b) 1.968   (c) 1.094
    (d) 0.031 4   (e) 0.581   (f) 0.925
4) (a) 852.8   (b) 39.28   (c) 2.179
5) 0.003 573   **6)** 0.016 42
7) 0.589 6   **8)** 131.8
9) 0.741 1   **10)** 44.24
11) 0.005 49   **12)** 0.769

ANSWERS TO CHAPTER 6

Exercise 10

1) 1, 3
2) (a) 1, 3   (b) $-3, 4$
    (c) $-31, -1.7$ (d) 4.3, -2.5
3) 2, 1   **4)** 3, 4
5) 4.74, 1.0   **6)** $a = 0.030\ b = 0$
7) 100, 0.43   **8)** 1.29, 20
9) $k = 0.02$ **10)** $m = -28.3, c = -42$

Exercise 11

1) $a = 2, b = 3$
2) $m = 1.5, c = 0.5$
3) $a = 0.1, b = 0.2$
4) $a = 3, b = 1$
5) $a = 0.2, b = 2$
6) $a = 70, b = 50$
7) $k = 0.2$
8) $a = 0.05, b = 5$
9) $k_1 = 0.1, k_2 = 0.006$
10) $a = 0.2, b = 0.1$

Self Test 2

1) d	**2)** b, d	**3)** b	**4)** none
5) b, c	**6)** d	**7)** c	**8)** d
9) b	**10)** b		

ANSWERS TO CHAPTER 7

Exercise 12

1) 1, 2	**2)** 4, 5	**3)** 4, 1
4) 4, 1	**5)** 7 3,	**6)** $5.23 - 1.35$

Exercise 13

1) 3, 4   **2)** 4 repeated
3) $+3, -3$ **4)** 3, -5
5) 0.667, 7   **6)** $-5, -1.5$
7) 2.414, -0.414 **8)** 2.181, 0.153
9) $+0.745, -0.745$

Self Test 3

1) b	**2)** d	**3)** e	**4)** a
5) c	**6)** b		

ANSWERS TO CHAPTER 8

Exercise 14

1) $a = 3, n = 2$
2) $a = 2 \times 10^{-6}, n = 4$
3) $n = 2, R = 10$
4) $e = 9.2(0.5)^t$
5) $k = 23.3, c = 2.99$
6) $V = 100, t = 0.002\ 5$

Exercise 15

1) $a = 3, n = 0.5$
2) $n = 4$, for $V = 80$ read $V = 70$

3) $t = 0.3m^{1\cdot5}$

4) $k = 100$, $n = -1.2$

5) $a = 245$, $b = 33$

6) $\mu = 0.5$, $k = 5$

7) $I = 0.02$, $T = 0.2$

ANSWERS TO CHAPTER 9

Exercise 16

1) (a) 1.35   (b) 12.18   (c) 90.02
   (d) 0.67   (e) 0.37   (f) 0.03

2) $y = 1.82$, $x = 0.84$

3) $x = -1.39$, -0.5

4) (a) $T = 631$ (b) $s = 1.12$

5) 4.34 mA per second.

6) (a) 0.18 seconds
   (b) 1200 volts per second.

ANSWERS TO CHAPTER 10

Exercise 17

1) -5, 19   2) 39.5, 5, 17

3) -4, 8   4) 3

5) 5.52, -6.52 6) 2.5, 2, 1

7) 2

Exercise 18

1) $2x$ 2) $7x^6$ 3) $12x^2$

4) $30x^4$ 5) $1.5t^2$ 6) $2\pi R$

7) $\frac{1}{2}x^{-1/2}$ 8) $6x^{1/2}$ 9) $x^{-1/2}$

10) $2x^{-1/3}$ 11) $-2x^{-3}$ 12) $-x^{-2}$

13) $-\frac{3}{5}x^{-2}$ 14) $-6x^{-4}$ 15) $-\frac{1}{2}x^{-3/2}$

16) $-\frac{1}{3}x^{-3/2}$ 17) $-\frac{15}{2}x^{-5/2}$ 18) $\frac{3}{10}t^{-1/2}$

19) $-0.01h^{-2}$ 20) $-5x^{-2}$ 21) $8x-3$

22) $9t^2-4t+5$ 23) $4u-1$

24) $20x^3-21x^2+6x-2$

25) $35t^4-6t$ 26) $\frac{1}{2}x^{-1/2}+\frac{5}{2}x^{3/2}$

27) $-3x^{-2}+1$ 28) $\frac{1}{2}x^{-1/2}-\frac{1}{2}x^{-3/2}$

29) $3x^2-\frac{3}{2}x^{-3/2}$

30) $1.3t^{0\cdot3}+0.575t^{-3\cdot3}$

31) $\frac{8}{5}x^2-\frac{4}{7}x-\frac{1}{2}x^{-1/2}$ 32) $-0.01x^{-2}$

33) $4.65x^{0\cdot5}-1.44x^{-0\cdot4}$

34) $\frac{3}{2}x^2+5x^{-2}$ 35) $-6+14t-6t^2$

36) -5, 19   37) 39.5, 5, 17

38) -4, 8   39) 3

40) 5.52, -6.52 41) 2.5, 2, 1

42) 2

Exercise 19

1) $4\cos x$ 2) $\cos x$

3) $1-4\cos x$ 4) $3\cos x$

5) $-3\sin\theta$ 6) $1-4\sin t$

7) $2(\theta+\sin\theta)$ 8) $-2\sin\theta+3\cos\theta$

9) $9.6t+1/4\sin t$ 10) $\dfrac{-10}{v^2}+6\cos v$

11) (a) $2-\sin x$ is always > 1
    (b) $\cos t+1$ is always $\geqslant 0$

12) (a) 1   (b) 0   (c) -1

Exercise 20

1) $5(x+1)^4$ 2) $9(3x+2)^2$

3) $4(6+x)^3$ 4) $28(4x-1)^6$

5) $-3(3-x)^2$ 6) $-9(2-3x)^2$

7) $-4(2-x)^3$ 8) $-10(4-2x)^4$

9) $8x(x^2-1)^3$ 10) $9x^2(1+x^3)^2$

11) $20x^3(x^4-2)^4$ 12) $-10x^4(1-x^5)$

13) $(2t+3)^{-1/2}$ 14) $(30-1)^{-2/3}$

15) $-4(r-1)^{-5}$ 16) $6(2-3t)^{-3}$

17) $(2h-5)^{-1/2}$ 18) $-2(1-4r)^{-1/2}$

19) $6(1-x)^{-2}$ 20) $-16(2x+7)^{-3}$

21) $4\cos 4x$ 22) $4\cos(4x+3)$

23) $10\pi\cos(2\pi t-0.04)$

24) $-2\sin 2x$ 25) $-2\sin(2x+3)$

26) $-4000\pi\sin(100\pi t)$

27) $-500\pi\sin(100\pi t-0.001)$

28) $-15\sin(3t-0.1)$

29) $500\pi\cos(2\pi t)$

30) $100\pi\sin(0.05-2\pi t)$

ANSWERS TO CHAPTER 11

Exercise 21

1) 20 mm   2) 3 m^2

3) 1400 cm^2 4) 630

Exercise 22

1) 752   2) 172   3) 0.8

4) 99   5) 25.5   6) 1095 m^2

Exercise 23

1) 0.64   2) 5   3) 1.67   4) 2.25

5) (a) 0.71 (b) 0.71   6) 5.77

7) 2.36   8) 2.45

Self Test 4

1) d   2) b   3) d   4) d

5) c   6) d   7) c   8) a

9) c   10) d

ANSWERS TO CHAPTER 12

Exercise 24

1) (a) $\{3,4,5,6,7,8,9,11\}$
   (b) $\{4,6,8\}$ (c) C
(d) $\{3,5,7\}$ (e) $\{2,4,6,8,9,10,11\}$
(f) $\{2,4,6,8,9,10,11\}$ (g) $\{2,10\}$
(h) $\{2,4,6,8,9,10,11\}$

2) (a) 0   (b) 1   (c) A (d) Y (e) X

4) (a) 0   (b) $X+Y(W+Z)$

Exercise 26

2) (a) $AB+A\bar{B}+\bar{A}\bar{B}$
   (b) $X\bar{Y}Z+X\bar{Y}\bar{Z}+\bar{X}YZ+\bar{X}\bar{Y}Z+XYZ$
   (c) $XYZ+\bar{X}YZ+XY\bar{Z}+X\bar{Y}Z+X\overline{YZ}$

3) $f_1 = Y\bar{Z}, f_2 = \bar{Y}Z+Y\bar{Z},$
   $f_3 = YZ+\bar{Y}\bar{Z}$

4) (a) $\bar{A}B$ (b) $\bar{X}\bar{Y}\bar{Z}+XY\bar{Z}+\bar{X}Y\bar{Z}$
   (c) $\bar{X}\bar{Y}\bar{Z}+\bar{X}\bar{Y}Z+\bar{X}Y\bar{Z}$

Exercise 27

1) (a) (b) (c)

2)
(a)

(b)

(c)

3) (a) $\bar{A}\bar{B}+AB$ (b) $\bar{X}\bar{Y}+YZ$ (c) $A\bar{C}+\bar{B}C$
   (d) $\bar{B}\bar{D}+BD$

4) (a) $X+\bar{Y}+Z$ (b) $A+BC$
   (c) $X+\bar{Y}+Z$ (d) $\bar{B}$ (e) $\bar{X}\bar{Y}$
   (f) $WX+\bar{W}Z$
   (g) $AD+\bar{B}D+CD+\bar{A}BC$

Exercise 28

1)
(a)   (b)

(c)   (d)

2) (a) $\bar{X}+\overline{YZ}$ (b) $\overline{X+YZ}$
   (c) $\overline{X\bar{Y}}+Z$

3)

Self-Test 5

1) c   **2)** b, c   **3)** a, d   **4)** a, d   **5)** a
6) b   **7)** b   **8)** d   **9)** c   **10)** a

Exercise 30

4) 4, 2.094 seconds   **5)** $3\sin 2t$

Exercise 31

4) (a) $\sin\left(\omega t+\dfrac{\pi}{2}\right)$ (c) $\sin \omega t$

   (b) $\sin(\omega t-\pi)$ (d) $\sin\left(\omega t-\dfrac{\pi}{6}\right)$

Exercise 32

1) (a) 0.615 7, 0.788 0
   (b) 0.955 1, 0.309 0
   (c) 0.615 7, $-0.788\,0$
   (d) 0.951 1, $-0.309\,0$
   (e) $-0.342\,0$, $-0.939\,7$
   (f) $-0.939\,7$, $-0.342\,0$
   (g) $-0.819\,2$, 0.573 6
   (h) $-0.529\,9$, 0.848 0

2) $30°, 150°; -2.82$

3) $78°, 282°$ **4)** $23°, 157°$

5) $v_R = 6\sin\theta$
6) $v_R = 7.7\sin(\theta-19°)$
7) $i_R = 6.6\sin(\theta+12°)$
8) $v_R = 51\sin(\theta-11°)$
9) $38\sin 5\pi t$, 12.3
10) 14.9, 6.7, 99.8

Exercise 33

1) (a) 1.573   (b) 1.083   (c) 1.019
    (d) 1.538   (e) 1.288   (f) 0.283 3
    (g) 0.023 9   (h) 0.848 6
    (i) 0.008 7   (j) 0.058 4
    (k) 0.074 3   (l) 0.475 9

2) (a) 47° 40′   (b) 57° 19′   (c) 63° 49′

3) (a) 8.007   (b) 6.145   (c) 9.396
    (d) 13.65   (e) 7.437   (f) 18.43

4) (a) 30° 55′   (b) 23° 10′   (c) 63° 42′

5) 76.03 mm   **6)** 42° 54′   **7)** 6.833 cm

9) 3.875 cm   **10)** 108.7 mm

Exercise 34

11) 0.334 6   **12)** 0.522 5, 1.632

13) 1.524, 0.656 1, 0.754 7

14) 1.110, 0.900 9

Exercise 35

1) 22.1 cm²   **2)** 31.7 cm²

3) 2765 mm each side

4) 11.9 cm²   **5)** 28 cm²

6) 89 cm²   **7)** 2.12 m

8) 3060 mm²

9) (a) 13.8 cm²   (b) 65 cm²

10) (a) 3.31 mm²   (b) 19.3 mm²

11) 15.7 cm

ANSWERS TO CHAPTER 14

Exercise 36

1) (a) $18+j10$ (b) $-j$ (c) $8-j10$

2) (a) $-1+j3$ (b) $-4-j3$ (c) $10-j3$

3) (a) $-9+j21$ (b) $-36-j32$
    (c) $-9+j40$ (d) 34   (e) $-21-j20$
    (f) $18-j30$ (g) $0.069-j0.172$
    (h) $-0.724+j0.690$
    (i) $-0.138-j0.655$
    (j) $0.644+j0.616$ (k) $3.5-j0.5$
    (l) $0.2+j0.6$

4) (a) 1, $j0.5$ (b) 0, $j5$ (c) -3, $j3$

5) (a) $-1\pm j$ (b) $\pm j3$

6) $0.634-j0.293$

7) (a) $3+j4$ (b) $0.4+j0.533$
    (c) $0.692+j2.538$

Exercise 37

2) Mod 5, Arg 53° 8′; Mod 5, Arg $-36°52$

3) 3.61, 146° 19′

4) 4.47, $-153°$ 26′

5) (a) $5\ \angle 36°\ 52'$ (b) $5\ \angle -53°\ 8'$
    (c) $4.24\ \angle 135°$
    (d) $2.24\ \angle -153°\ 26'$
    (e) $4\ \angle 90°$ (f) $3.5\ \angle -90°$

6) (a) $2.12+j2.12$ (b) $-4.49+j2.19$
    (c) $4.32-j1.57$ (d) $-1.60-j2.77$

7) (a) $56\ \angle 70°$ (b) $10\ \angle -50°$
    (c) $15\ \angle 90°$ (d) $21\ \angle -90°$

8) (a) $2.67\ \angle -30°$ (b) $2\ \angle -60°$
    (c) $0.6\ \angle -9°$ (d) $2.83\ \angle 44°\ 39'$

9) (a) $30\ \angle 33°$ (b) $-2.93+j6.90$

10) (a) $0.277\ \angle 56°\ 19'$
    (b) $13\ \angle -112°\ 38'$

11) (a) $0.172\ \angle -59°\ 2'$
    (b) $0.0427-j0.0385$

12) (a) $r=4.5$, $X_L=2.2$
    (b) $R=23$, $X_C=35$
    (c) $R=27.2$, $X_L=11.7$
    (d) $R=6.85$, $X_C=1.46$

13) (a) 14° 37′   (b) 345 watts.

ANSWERS TO CHAPTER 15

Exercise 38

1) $18x$, 54   **2)** -187, 254

3) -20 **4)** 143.2

5) 15, 0, -135 **6)** -89

7) -0.115 **8)** 51 583

Exercise 39

1) 42 m/s   **2)** 6 m/s²

3) (a) 6 m/s   (b) 2.41 or -0.41 s
    (c) 6 m/s²   (d) 1 s

4) -0.074 m/s, 0.074 m/s²

5) 10 m/s, 30 m/s   **6)** 3.46 m/s

7) (a) 4 rad/s   (b) 36 rad/s²
    (c) 0 s or 1 s

8) (a) -2.97 rad/s   (b) 0.280 s
    (c) -8.98 rad/s²   (d) 1.57 s

9) 62.5 kJ

Exercise 40

1) (a) 11 (max), -16 (min)
    (b) 4 (max), 0 (min)
    (c) 0 (max), -32 (min)

2) (a) 54   (b) $x=2.5$ (c) $x=-2$

3) $(3, -15)$, $(-1, 17)$

4) (a) -2 (b) 1   (c) 9

5) (a) 12   (b) 12.48

6) 15 mm   **7)** 10 m   **8)** 4

9) 108 000 mm³

10) radius = height = 4.57 m

11) 405 mm

12) diameter = 28.9 mm, height 14.4 mm

13) 5.76 m

Exercise 41

1) Min(−0.25, −6.125)

2) Min(0, 0), Max(−1, 1)

3) Min(0, −4), Max(−3, 23)

4) Inf(0, 0), Max(3, 27)

5) Max(0, −2), Min(1, −3),
Min(−1, −3)

6) Max$\left(\dfrac{\pi}{4}, \sqrt{2}\right)$, Min$\left(\dfrac{5\pi}{4}, -\sqrt{2}\right)$

7) Max$\left(\dfrac{\pi}{3}, 2\right)$, Min$\left(\dfrac{4\pi}{3}, -2\right)$

INDEX